Chemistry

Bob Wilson
Assistant Headteacher,
Forrester High School, Edinburgh

Published by BBC Educational Publishing,
BBC White City, 201 Wood Lane, London W12 7TS
First published 2000, Reprinted 2001, 2002
© Bob Wilson/BBC Education 2000

ISBN: 0 563 47491 2

Designed by Linda Reed and Associates
Illustrations by Tech Type
Printed in Great Britain by Bell & Bain Ltd., Glasgow

Contents

Introduction 4

Chemical reactions and the periodic table 8
Chemical reactions 10
Speeding up chemical reactions 12
The periodic table 14
The atom 16

Bonding and properties of substances 20
Bonding 22
Properties of substances 24
Electrolysis 26

Fuels 28
Fossil fuels 30
Fractional distillation 32
Pollution 34
Hydrocarbons 36

Acids and alkalis 38
Acids and alkalis 40
Reactions of acids 42

Making electricity 46
Batteries and cells 48

Metals and corrosion — 52

Properties of metals/alloys — 54

Extracting metals — 56

Reactions of metals — 58

Corrosion — 60

Protecting iron — 62

Plastics — 64

Making plastics — 66

Using plastics — 70

Fertilisers — 72

The need for fertilisers — 74

Ammonia — 76

Nitric acid — 77

Carbohydrates and related compounds — 78

Photosynthesis and respiration — 80

Carbohydrates — 82

Making and breaking carbohydrates — 84

Alcoholic drinks — 85

Formulae, equations and calculations — 86

Answers — 94

Index — 96

Introduction

About BITESIZE Chemistry

Standard Grade BITESIZE Chemistry is a revision scheme to help achieve success in your Standard Grade (SG) exam. It consists of this book which covers all of the topics, a two-hour TV programme covering a number of topics in detail and an Internet site which includes 'Ask a teacher' where your particular questions are answered by a Chemistry teacher.

It is called BITESIZE because that's a good way to revise – in small chunks. All three elements of BITESIZE are divided into a number of sections so that you can go through them one by one, or select sections with which you are having difficulty. BITESIZE is a revision scheme and should be used in addition to your class notes. The TV programme can be taped at home or in school and if you're not online at home you can always access the website at school.

About this book

This book is divided into ten sections covering all of the topics and learning outcomes in the Standard Grade course.

Section	Standard Grade topics
Chemical reactions and the periodic table 📺	1. Chemical reactions
	2. Speed of reactions
	3. Atoms and the periodic table 🌐
Bonding and properties of substances	4. How atoms combine
	7. Properties of substances 🌐
Fuels 📺	5. Fuels 🌐
	6. Structures and reactions of hydrocarbons 🌐
Acids and alkalis 📺	8. Acids and alkalis 🌐
	9. Reactions of acids 🌐
Making electricity	10. Making electricity 🌐
Metals and corrosion 📺	11. Metals 🌐
	12. Corrosion
Plastics 📺	13. Plastics and synthetic fibres 🌐
Fertilisers 📺	14. Fertilisers
Carbohydrates	15. Carbohydrates and related substances
Formulae, equations and calculations	Formulae, equations and calculations 🌐

4

THE ONLINE SERVICE
🌐 You can find extra support, tips and answers to your exam queries on the Standard Grade BITESIZE website. The address is www.bbc.co.uk/scotland/revision

KEY TO SYMBOLS

Ⓒ Credit Level only

(?) A question to think about

◎ An activity to do

📺 A link to the video

🌐 A link to the website

Although formulae, equations and calculations are spread throughout the Standard Grade course they are dealt with in one special reference section at the back of the book (pages 86-93) to make it easier for you to find and revise.

Each section begins with an introductory page to tell you what the section is all about. This summarises the main ideas in the section. The next page is the FactZone which lists what you should know by the time you have worked through the section. It also shows which information is relevant to General and Credit Level. The rest of the sections give you more details about particular topics. The start of the Credit Level text is clearly marked with C. All other text is at General Level and is not marked with any symbol. Each sections includes activities and questions including past paper questions at General and Credit Levels to give you practice. Again, the level is clearly marked. The answers to the questions are located at the back of the book.

Throughout the book it is made clear where there is a link to the TV programme TV and Internet site Ⓞ.

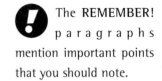 The **REMEMBER!** p a r a g r a p h s mention important points that you should note.

Revision tips

Everybody has their own preferred way of studying, but here are some useful tips for revising.

- Work in a quiet area. If you can't work at home, go to the school or public library.

- Plan your revision. Do a section at a time then have a break. Don't try to do the whole book in one massive session.

- Do each of the activities as you come to them and do the practice questions once you have revised the section.

- 'Look, cover, write, check' is a good way to learn things.

- Check the FactZone to make sure you have covered all you need to know. Tick each fact once you know and understand it.

- Make a set of flashcards with facts you are having difficulty remembering and look at them regularly on your own or with a friend.

Problem solving

BITESIZE focuses on knowledge and understanding, but it will also help with problem solving. In Standard Grade Chemistry problem solving involves you handling and processing information, making predictions and drawing conclusions.

When doing these activities, you will often by applying knowledge and understanding to solve problems, sometimes in unfamiliar situations. Don't panic if you read a question and it doesn't seem familiar –it's likely to be a problem solving question.

Here is a more detailed summary of the sort of things you have to be able to do in problem solving:

- select information from the SQA Data Booklet, graphs and diagrams
- present information as a table
- draw bar and line graphs with correctly labelled axes and units
- suggest the best way to carry out an experiment
- carry out a chemical test in a fair way and identify factors which could affect the fairness of a test
- draw a conclusion from information given
- explain results and conclusions
- make predictions and generalisations from information given.

The exam papers

There are separate General and Credit papers each lasting one and a half hours with a break in between. Most pupils sit both papers. Each paper has a total of 60 marks available, with approximately 30 marks for knowledge and understanding (KU) and 30 marks for problem solving (PS). The right-hand margin indicates whether a question is KU or PS. You write your answers on the question papers.

The first part of each paper has grid questions worth 20 marks. In a grid question, you choose the correct answer(s) from a grid containing between four and six possible answers. Look carefully at the way the question is asked – it indicates the number of correct answers required.

For example:

- 'Identify the symbol for...'

This indicates one answer only. If you give more than one answer you won't get any marks.

- 'Identify the two compounds which...'

This clearly indicates that two answers are needed to get 1 mark. If you give only one answer or more than two answers you will get no marks.

- 'Identify the correct statement(s) about ...'

This indicates that there could be more than one answer, but there will not be more than two answers.

The other 40 marks on the papers are for questions requiring short answers but some questions require extended (longer) answers.

Remember to read the inside cover of the exam paper carefully.

Picking up marks in exams!

Follow these tips to make sure you get all the marks you deserve.

- Read all the questions carefully.

- Make sure you answer the question asked.

- In grid questions, make sure you choose from the answers provided and don't leave any question unanswered.

- If you are asked to 'explain a trend', you must say what the trend is then explain it.

- Graphs – make sure you draw the type of graph asked for. As a general rule, two sets of numbers should be drawn as a line graph. Label the axes with units. Plot each point clearly. Draw a line or smooth curve through the points.

- Chemical formulae and equations – use the rules for working out formulae. Remember that an equation is simply the formulae for reactants and products put together.

- Calculations – you must show your working and units. Even if you can't do the arithmetic, you will get some marks for using the correct method and units.

Revision timetable

Fill in the revision timetable below to help you work through BITESIZE Chemistry and ensure that you have revised all the necessary subjects.

Date	Section covered	Time spent	Completed

Good luck!

Chemical reactions and the periodic table

⊛ This section is about:

- chemical reactions

- speeding up chemical reactions

- the atom.

Chemical reactions are happening around us all the time. Some reactions happen naturally – we cause others to happen. When a reaction occurs, new substances are formed. We can tell if a reaction takes place because changes occur: a substance changes the way it looks or an energy change takes place. Elements and compounds are involved in reactions.

Chemical reactions can happen whether the reactants are solid, liquid or gas. They also occur in solution. A solution is formed when a solute dissolves in a solvent. Symbols can be used to show which state a substance is in. Mixtures can be separated using a number of methods.

Chemical reactions occur at different speeds. The particle size, concentration and temperature of reactants affect the speed of a reaction. Catalysts can be added to speed up reactions. Many important reactions in our body and in industry rely on catalysts.

There are over 100 known elements, arranged in the periodic table so that those with similar properties are grouped together. Each element has a name and a symbol. All elements are made from atoms. Atoms themselves consist of protons, electrons and neutrons. The number of these particles varies for each element. The electrons are arranged in energy levels. Each element has a mass number and an atomic number. The atomic number identifies the element. Not all atoms in an element have the same number of neutrons – these are isotopes.

FactZONE

g When you have finished this section, you should know the following at General Level:

- One or more substances are formed during a chemical reaction. A change of appearance or an energy change can indicate that a reaction has taken place. Chemical reactions take place in everyday life, e.g. cooking food, burning gas, iron rusting.

- An equation can be written to summarise what's happening in a chemical reaction.

Reactants → Products

- Compounds are formed when elements react.

- A mixture occurs when substances are brought together but don't react.

- A solution is formed when a solute dissolves in a solvent.

- The speed of reaction can be increased by using small particles, high concentrations of reactants and by heating the reactants. Catalysts speed up reactions but are not used up and are unchanged at the end of the reaction. Catalysts are used in everyday life, e.g. catalytic converters in car exhaust systems.

- All matter is made up from the one hundred or so elements. Elements are arranged in the periodic table and each has a name and a symbol. Elements can be classified as metal/non-metal; solid, liquid or gas; or natural/man-made.

- A group is a vertical column of elements and a period is a horizontal row. Elements in the same group of the periodic table show similar chemical properties, e.g. the alkali metals, the halogens, the noble gases and the transition metals.

- The alkali metals (group 1) are all very reactive. The noble gases (group 0) are all very unreactive.

- Every element is made up of atoms. The atom has a nucleus containing positively charged protons with negatively charged electrons moving around this nucleus. An atom is neutral because the positive charge is balanced by the negative charge.

- All elements have an atomic number.

- The electron arrangement for elements can be found in the SQA Data Booklet. Elements with the same number of electrons in the outer energy level are in the same group and have similar chemical properties.

C In addition, you should know the following at Credit Level:

- The nucleus of an atom also contains neutrons.

- A proton has a mass of 1 and a 1+ charge. An electron has a mass of practically 0 and a 1- charge. A neutron has a mass of 1 and no charge.

- The number of protons and electrons in an atom are equal. The atomic number is the same as the number of protons.

- The electrons are arranged in energy levels.

- The mass number = protons + neutrons, and neutrons = mass number – protons.

- Mass number and atomic number can be shown in nuclide notation, e.g. $^{35}_{17}Cl$. (mass number, atomic number)

- During chemical reactions, atoms can gain or lose electrons to form either positive or negative ions. Nuclide notation can be written for ions, e.g. for a chloride ion (Cl^-): $^{35}_{17}Cl^-$

- Most elements exist as a mixture of isotopes. Relative atomic mass is the average mass of the isotopes of an element.

- Relative atomic masses (RAM) are found on page 4 of the SQA Data Booklet.

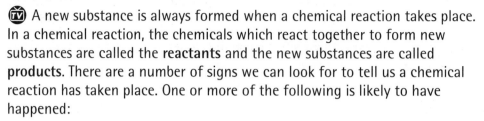

Chemical reactions

📺 A new substance is always formed when a chemical reaction takes place. In a chemical reaction, the chemicals which react together to form new substances are called the **reactants** and the new substances are called **products**. There are a number of signs we can look for to tell us a chemical reaction has taken place. One or more of the following is likely to have happened:

- there is a change in colour
- a gas is given off
- there is a change in temperature.

We can see these sorts of changes taking place in everyday life:

when baking a cake the ingredients are like chemical reactants which join together to make a new substance – the cake.

ingredients (reactants) → cake (product)

This is a word equation – a shorthand way of telling you what's happening in a reaction.

Other everyday reactions are shown below.

Very slow	Slow	Fast	Very fast
Car body rusting	Baking a cake	Gas burning	Explosion

Chemical reactions are also carried out in the laboratory. In each of the examples below, some kind of change takes place to indicate a reaction has taken place:

magnesium	+	oxygen	→	magnesium oxide
(grey metal)		(colourless)		(white powder)
element		element		compound

A bright flame is seen and lots of heat is given out.

Compounds can also react with elements or other compounds to form new compounds.

! **R E M E M B E R** An **element** is made up of one kind of atom.

Acid + chalk

! **R E M E M B E R** A **compound** is made up of two or more elements joined together.

The bubbles of gas show a reaction is taking place.

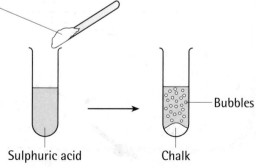

Chalk

Bubbles

Sulphuric acid   Chalk

Acid + alkali

hydrochloric acid + sodium hydroxide → sodium chloride + water
(colourless   (colourless   (colourless   (colourless
solution)   solution)   solution)   liquid)

No change in appearance occurs but a rise in temperature indicates a reaction is taking place.

Solutions

Many chemical reactions take place in **solution**. A solution is formed when a **solute** dissolves in a **solvent**.

The **solute** is the substance being dissolved.

The **solvent** is the liquid doing the dissolving.

> **!** R E M E M B E R
> A mixture is formed when two or more substances come together but do not react.

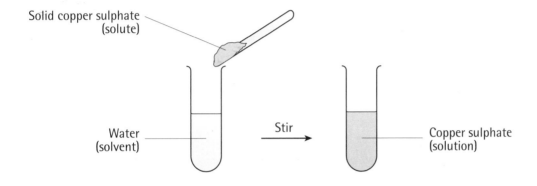

Solid copper sulphate (solute)

Water (solvent)   Stir →   Copper sulphate (solution)

Practice questions

1   Two colourless solutions were mixed in a beaker and no change was seen to have taken place. However, a pupil said a reaction had occurred.

   a) What change could have taken place to make the pupil realise a reaction had occurred?

   b) How could she prove that this change had taken place?

2   Chemical reactions happen all around us.

A		B	
	water boiling		paint drying
C		D	
	gas burning		iron rusting

Identify two chemical reactions from A–D.

3   When salt is added to water and stirred, it dissolves.
Identify the solute and solvent.

4   When grey magnesium metal is added to a solution of blue copper(II) sulphate, a colourless solution of magnesium sulphate and copper metal are formed.

   a) What would you see happening which would indicate that a chemical reaction had occurred?

   b) Write a word equation for this reaction.

   c) Which substances are elements?

Speeding up chemical reactions

📺 You can speed up a chemical reaction if you:

- increase the temperature of the reactants
- reduce the particle size of the reactants
- increase the concentration of the reactants
- add a catalyst.

This can be shown in everyday reactions:

- when baking, heat is used to increase the speed at which a cake is made
- metal filings (small particles) are used in fireworks so that they burn quickly and easily
- when preparing a barbecue, fanning the coals with a piece of card makes them burn faster because the concentration of oxygen is increased.

Experiments can be carried out in the laboratory to show these effects.

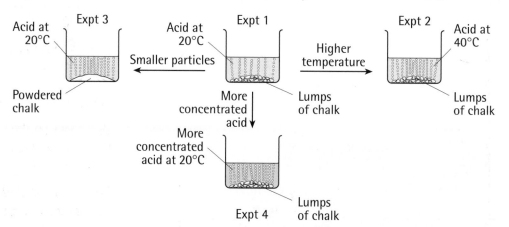

The gas can be collected and a graph of volume of gas against time can be drawn.

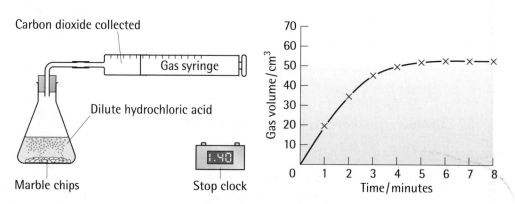

The steeper the slope of the graph, the faster the reaction.

◎ *Add a dotted line to the graph above to show the graph you would get when the reaction was carried out at a higher temperature.*

Catalysts

Catalysts speed up reactions but are not used up during the reaction. They remain chemically unchanged at the end of the reaction. A catalytic converter in a car exhaust system quickly changes harmful gases into harmless ones. Metals like iron and platinum are used in industry to speed up the manufacture of ammonia and nitric acid. Enzymes are natural catalysts. Our body has enzymes which make essential chemical reactions happen quickly.

Practice questions

1   George investigated the reaction of magnesium and dilute sulphuric acid. Here are his notes.

At first the reaction was slow.
The flask got very hot.
The rate of the reaction speeded up.
The reading on the balance fell.
After five minutes the reaction had stopped.

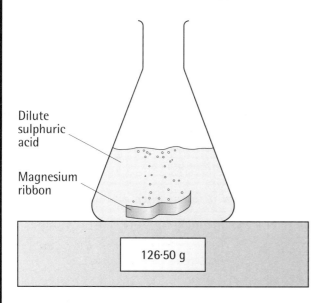

Dilute sulphuric acid

Magnesium ribbon

126·50 g

Magnesium + sulphuric acid ⟶ magnesium sulphate + hydrogen

a) Suggest why the rate of reaction speeded up.

b) Why did the reading on the balance fall during the reaction?

2   When Stephanie added manganese dioxide to hydrogen peroxide solution, oxygen was produced.

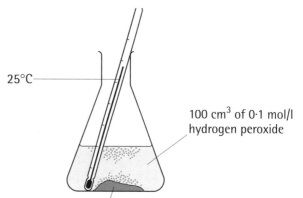

25°C

100 cm^3 of 0·1 mol/l hydrogen peroxide

1 g manganese dioxide

a) What is the purpose of a catalyst?

b) What will be the mass of the manganese dioxide at the end of the reaction?

c) Stephanie wanted to see if raising the temperature to 35°C would speed up the reaction.

Complete the labelling of the diagram to show how she would make her second experiment a fair test.

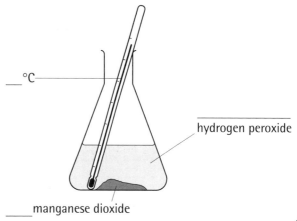

—°C

hydrogen peroxide

manganese dioxide

The periodic table

📺 Elements

An element is a pure substance made up of only one kind of atom. It can't be broken down into anything simpler. There are more than 100 known elements, each with a name and symbol. They are all arranged in the periodic table according to their properties. The elements are arranged in **groups** and **periods**. Groups run vertically, periods horizontally.

When you're thinking about the periodic table look for patterns in the way it is laid out.

Group 1 – the alkali metals

The alkali metals are very reactive. Lithium, sodium and potassium are so reactive they are stored under oil to stop them reacting with oxygen in the air. Rubidium and caesium are so reactive they are kept in a vacuum. They are all soft metals, easily cut. They quickly lose their shine because they react so quickly in air. They all react violently with water and often burst into flames – each metal has its own flame colour. Hydrogen gas is always produced and an alkaline solution is formed.

REMEMBER
The state of each reactant and product is given by the state symbols in brackets: (s) = solid, (l) =liquid, (g) = gas and (aq) = aqueous (solution). The reactivity of the metals increases as you move down the group.

Word equation: sodium + water → hydrogen + sodium hydroxide

Formula equation: $Na(s)$ + H_2O (l) → $H_2(g)$ + $NaOH(aq)$

Balanced equation: $2Na(s)$ + $2H_2O$ (l) → $H_2(g)$ + $2NaOH(aq)$

📺 ◎ *Watch the section of the video which covers the reaction of alkali metals with water and complete the table below.*

Element	Symbol	How it reacts with water
Lithium	Li	Floats and fizzes; gives off hydrogen
Sodium		
Potassium		
Rubidium		
Caesium		

REMEMBER
Make sure you know some of the properties of the alkali metals, noble gases, halogens and transition metals.

Group 0 – the noble gases

The noble gases are all very unreactive. This property can be put to good use. Helium is less dense than air and it doesn't burn. It is therefore used in weather balloons and airships. Argon is used in light bulbs because it will not react with the hot filament.

Group 7 – the halogens

The halogens are the most reactive of the non-metals. They are only found as compounds because the elements are so reactive. Sodium chloride is the chemical name of common table salt. Fluoride compounds are added to water and toothpaste to help protect teeth from decay.

The transition metals

The transition metals form a large grouping in the middle of the periodic table. Although they are not in one group they still show many similarities. They are generally unreactive but those which do react tend to form coloured compounds. Silver and gold are precious metals. Emeralds and sapphires are jewels containing transition metal compounds. Many transition metals are used as catalysts, e.g. platinum and rhodium in catalytic converters, and iron in the manufacture of ammonia.

> **! REMEMBER**
> You don't have to memorise any of the groups in the periodic table – they are all found in the SQA Data Booklet.

Practice questions

1   From the periodic table, pick an element which fits each of the descriptions below and write down its symbol.

   a) A metal which is soft and reacts very fast with water.

   b) A metal used as a catalyst in car exhausts.

   c) An element which forms compounds that are added to toothpaste to help stop tooth decay.

   d) A gas which can be used in light bulbs.

   e) The first element in group 2.

   f) The second element in the third period.

   g) A metal which is a liquid at room temperature.

2

A	B	C
chlorine	copper	oxygen
D	E	F
lithium	sulphur	bromine

   a) From the grid, identify the **two** elements in the same group as fluorine.

   b) Identify the element which is a transition metal.

   c) Identify the two elements which were discovered in 1774.

3

A	B	C
Ca	Cd	Ce
D	E	F
Cl	Cr	Cs

   a) From the grid, identify the symbol for cadmium.

   b) Identify the symbol for an alkali metal.

The atom

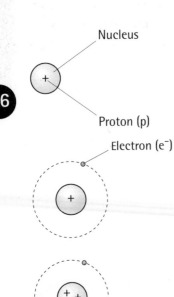

Nucleus

Proton (p)

Electron (e⁻)

📺 Everything in the world is made of **atoms**.

Every atom has a **nucleus** at its centre containing **positively** charged particles called **protons** (p).

Negatively charged **electrons** (e⁻) move around the nucleus.

Atoms have a neutral charge overall because there are the same number of protons as electrons, which have equal but opposite charges.

Their charges balance each other, e.g. each lithium atom has three protons and three electrons so the overall charge is neutral.

Ⓒ   Atoms also have particles called **neutrons** (n) in the nucleus. Neutrons, as their name suggests, are neutral – they have no charge. Neutrons have a **mass of 1**. Protons have a **mass of 1** and a **1+** (one positive) charge. **Electrons** have practically **no mass** and a **1-** (one negative) charge.

📺 ◎ *Complete this table to summarise the structure of the atom.*

Particle	Symbol	Charge	Mass	Location
proton electron neutron				

Atomic number

❗ **R E M E M B E R** All the elements are different because their atoms have different numbers of protons, electrons and neutrons.

Although all elements are made of atoms, the elements are all different. This is because the atoms of each element are different. The difference lies in the number of particles each element has.

The elements are arranged in order of increasing **atomic number**. The atomic number is the same as the number of **protons**. Because atoms contain the same number of protons as electrons, the atomic number also tells us how many electrons are in an atom. As the atomic number increases the number of particles in the atom increases. This means the size and mass of the atoms increase.

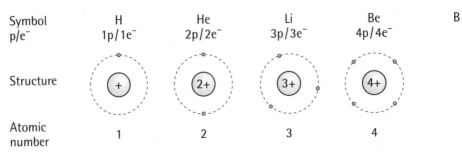

Symbol	H	He	Li	Be	B
p/e⁻	1p/1e⁻	2p/2e⁻	3p/3e⁻	4p/4e⁻	
Structure	+	2+	3+	4+	
Atomic number	1	2	3	4	

◎ *Draw the structure of the boron atom in the space above and write down the number of protons and electrons and the atomic number.*

Electron energy levels

The electrons in an atom are arranged in energy levels – also known as shells. There is a maximum number of electrons in each energy level.

Example: sodium (Na): 2, 8, 1

This means each sodium atom has 2 electrons in the first energy level, 8 in the second and one in the outer energy level.

◉ *Look up the electron arrangements of chlorine and calcium in the SQA Data Booklet.*

Notice that the total number of electrons is the same as the atomic number. Also note that the number of electrons in the outer energy level is the same as the group number. This means elements with the same number of electrons in the outer energy level of each atom have similar chemical properties.

◉ *Look up the electron arrangements of the alkali metals, halogens and noble gases. Look particularly at the number of outer electrons and the group number.*

G 📺 The mass number

The **mass number** is the number of protons and neutrons added together – they are the particles in an atom which have mass.

mass number = p + n

From this we can calculate the number of neutrons in an atom:

n = mass number – p

This is the same as saying:

n = mass number – atomic number

Example: An atom of sodium has 12 neutrons. Calculate the mass number.

Answer: Sodium has an atomic number of 11 so has 11 protons.

mass number = p + n

= 11 + 12

mass number = 23

The mass number and atomic number of an atom of an element can be shown as:

$^{\text{mass number}}_{\text{atomic number}} X$ where X is the element symbol.

e.g. $^{23}_{11}Na$

This is known as nuclide notation.

◉ *Calculate the mass number of an atom of lithium which has four neutrons. Write the nuclide notation for this atom.*

REMEMBER
You don't have to memorise the electron arrangements of elements – they are in the SQA Data Booklet.

BITESIZEchemistry

Given the nuclide notation, the number of protons, electrons and neutrons can be calculated.

Example: $_{17}^{35}Cl$: $p = 17$, $e^- = 17$, $n = 35 - 17 = 18$.

◎ *Calculate the number of protons, electrons and neutrons in an atom of magnesium with the nuclide notation $_{12}^{25}Mg$.*

It is possible for atoms to lose and gain electrons during chemical reactions. This results in charged atoms called **ions**. The nuclide notation will be the same but the symbol will now carry a charge.

A sodium atom loses one electron when it reacts and forms a 1+ ion, i.e. Na^+.
The nuclide notation for the ion is $_{11}^{23}Na^+$.

The number of protons and neutrons is the same as the atom – only the number of electrons is different.

$_{11}^{23}Na$ atom	$_{11}^{23}Na^+$ ion	$_{17}^{35}Cl$ atom	$_{17}^{35}Cl^-$ ion
$p = 11$	$p = 11$	$p = 17$	$p = 17$
$n = 12$	$n = 12$	$n = 18$	$n = 18$
$e^- = 11$	$e^- = 10$	$e^- = 17$	$e^- = 18$

◎ *Calculate the number of protons, neutrons and electrons in a magnesium ion with the nuclide notation $_{12}^{25}Mg^{2+}$.*

G Isotopes

Isotopes are atoms of the same element but with different numbers of neutrons, i.e. different mass numbers. They have the same number of protons so are from the same element. Most elements exist as mixtures of isotopes.

Chlorine has two isotopes:

! REMEMBER
Isotopes are atoms with the same atomic number but different mass numbers.

$_{17}^{35}Cl$	$_{17}^{37}Cl$
$p = 17$	$p = 17$
$e^- = 17$	$e^- = 17$
$n = 35 - 17 = 18$	$n = 37 - 17 = 20$

◎ *Hydrogen has three isotopes: $_1^1H$, $_1^2H$ and $_1^3H$. Calculate the number of neutrons in each.*

◉ Relative atomic mass (RAM)

The total mass of an atom comes from the mass of its neutrons and protons. The electrons are so light that they don't really contribute to the mass. Most elements, however, have isotopes. Therefore, an average is taken of the mass of all the isotopes and this average mass is called the **relative atomic mass**.

The relative atomic mass of chlorine is 35.5. You might expect the average to be 36 but there are more ^{35}Cl atoms in any sample of chlorine gas so the average is nearer 35 than 37.

◎ *Look up relative atomic masses in the SQA Data Booklet. Notice that chlorine is not the only element which has a relative atomic mass which is not a whole number.*

⊘ ◉ *Bromine has two isotopes $^{79}_{35}Br$ and $^{81}_{35}Br$ and its relative atomic mass is 80. What does this tell you about the percentage of each isotope in bromine?*

Practice questions

1   The table gives the numbers of protons and neutrons in the nuclei of atoms of different elements.

Element	Number of protons	Number of neutrons
boron	5	6
phosphorous	15	16
zinc	30	35
zirconium	40	51
tin	50	69

a) The nucleus of an atom has a positive charge. Why is an atom neutral?

b) From the information in the table, write a statement linking the number of protons and neutrons in atoms.

◉ 2   There are three different types of silicon atom.

Type of atom	Number of protons	Number of neutrons
$^{28}_{14}Si$		
$^{29}_{14}Si$		
$^{30}_{14}Si$		

a) Complete the table to show the number of protons and neutrons in each type of silicon atom.

b) What name is used to describe these different types of silicon atom?

Bonding and properties of substances

⚙ This section is about:

- bonding – how atoms combine

- properties of substances

- electrolysis.

Atoms of elements can combine in a number of ways. Non-metal elements usually combine to form molecules. Molecules are groups of atoms held together by covalent bonds. Both elements and compounds can exist as covalent molecules. The smallest molecules are made up of two atoms – they are said to be diatomic. Larger molecules may contain thousands of atoms. Molecules have a variety of shapes.

Covalent molecular substances tend to be gases, liquids or solids with low melting and boiling points because molecules are easy to separate from each other. Some covalent substances, however, are solids with high melting and boiling points because they are giant networks and it takes a lot of energy to break covalent bonds. Most covalent substances are insoluble in water but do dissolve in other solvents.

Metals and non-metals form ionic bonds when they react. The metal atom transfers electrons to the non-metal atom so that both have a stable arrangement of electrons. The metal forms a positive ion and the non-metal forms a negative ion. The oppositely charged ions attract and form a giant crystal lattice. Ionic compounds have high melting and boiling points because it takes a large amount of energy to separate the ions.

When an ionic substance dissolves in water, the ionic lattice breaks up completely.

Electrical conductivity is a flow of electrical charge and is a property shown by metals and ionic compounds but not by covalent substances. In metals the flow of charge is moving electrons. In ionic compounds the flow of charge is moving ions. Conductivity can only take place when the compound is in solution or has been melted because the ions are then free to move. In the solid state ions are held tightly in the lattice and are not free to move. Covalent substances do not conduct electricity in any state as they have no electrons free to move and they are not made up of ions.

Electrolysis is the breaking up of a compound by passing electricity through it. Only ionic compounds can be electrolysed either in solution or as a melt. When a direct current (d.c.) is passed through electrodes, the ions are attracted to the oppositely charged electrode. Metal ions form metal atoms at the negative electrode and non-metal ions form the non-metal element at the positive electrode. Many ionic compounds are coloured because their ions are coloured. Coloured ions can be seen moving towards electrodes in migration experiments.

ⓖ When you have finished this section, you should know the following at General Level:

■ Atoms can be held together by bonds. A covalent bond is a shared pair of electrons. A molecule is a group of atoms held together by covalent bonds.

■ The molecular formula tells you the number of atoms of each element in a molecule. Only non-metal elements bond to form molecules. A diatomic molecule is made up of two atoms. Hydrogen, nitrogen, oxygen and the halogens are elements which exist as diatomic molecules.

■ Covalent substances usually exist as gases, liquids or low melting point solids, but some exist as networks.

■ Ionic bonds are formed between metals and non-metals. Metals lose electrons to form positive ions and non-metals gain electrons to form negative ions. Ionic compounds exist as lattices of oppositely charged ions.

■ An electric current is a flow of charged particles. Electrons flow through metals and graphite (carbon). Ions flow through ionic solutions or melts. An ionic lattice breaks up completely when dissolved in water. Ionic solids do not conduct electricity because the ions cannot move.

■ Covalent substances do not conduct electricity in any state. Wax is an example of a covalent substance which does not dissolve in water but does dissolve in a covalent solvent like hexane.

■ Electrolysis is the breaking down of a compound by passing a direct current through it. An electrolyte is a solution or melt which conducts electricity.

■ During electrolysis, positive metal ions are attracted to the negative electrode and a metal is formed while negative non-metal ions are attracted to the positive electrode and a non-metal element is formed.

■ Many ionic compounds are coloured because their ions are coloured. Migration experiments can be carried out to show coloured ions moving towards the electrodes.

ⓒ In addition, you should know the following at Credit Level:

■ Atoms bond to achieve a stable electron arrangement. In a covalent bond, two positively charged nuclei are held together by their common attraction for the shared pair of electrons.

■ You should be able to draw diagrams to show how outer electrons form a covalent bond and the shapes of two-element molecules.

■ Ionic compounds are solid because a lot of energy is needed to separate the ions. Covalent networks are solid because a lot of energy is needed to break the covalent bonds. Covalent molecular substances have low melting and boiling points because it does not take a lot of energy to separate molecules.

■ A direct current (d.c.) must be used in electrolysis in order to identify products at electrodes. At the negative electrode, positive metal ions gain electrons while at the positive electrode, negative non-metal ions lose electrons. Only ionic compounds can be electrolysed because only they are made up of ions.

■ Ion–electron equations are found in the SQA Data Booklet.

Bonding

When atoms join together they are said to have bonded. The two main types of bonding are **covalent** and **ionic**. Atoms bond in order to become stable – they end up with the same number of electrons in their **outer energy level** as a noble gas, i.e. two or eight.

22

! REMEMBER Generally, the number of atoms in a molecule indicates its shape: two – linear, three – planar, four – pyramidal and five – tetrahedral. You should be able to draw these shapes.

Covalent bonding

Atoms **share electrons** in order to obtain a stable arrangement of electrons. A single covalent bond is formed when two outer energy electrons are shared. A group of atoms held together by covalent bonds is called a **molecule**. Molecules are usually formed between non-metal atoms. Covalent bonds can form between atoms of the same element and atoms of different elements. The molecular formula of a substance indicates whether it is covalently bonded and how many atoms are in the molecule.

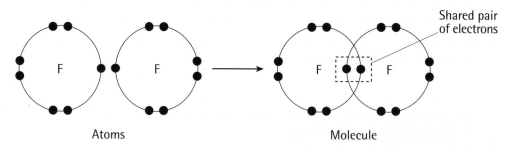

Atoms   Molecule

Shared pair of electrons

Molecule	Arrangement of atoms	Shape
Hydrogen chloride	H — Cl	Linear
Hydrogen oxide (water)	H $\diagup$ O $\diagdown$ H	Planar (flat)
Nitrogen hydride (ammonia)	H H H — N	Pyramidal
Carbon hydride (methane)	H — C — H H H	Tetrahedral

C When a covalent bond is formed, it is the common attraction of the positive nuclei for the shared pair of electrons which holds the atoms together. The way atoms bond results in molecules having special shapes.

◎ *Draw the shape of the molecule with the formula XY$_3$.*

Ionic bonding

Ionic bonds are formed between metals and non-metals. Metals transfer electrons to non-metals and, as a result, metals form positive ions and non-metals form negative ions. The **oppositely charged ions attract** each other and form a **lattice**. Both the positive and negative ions have a stable arrangement of electrons. The charge on the ion depends on the number of electrons lost and gained.

Example 1:

Sodium atoms reacting with chlorine atoms to form sodium chloride.

Atoms	Electron arrangement	Ion	Electron arrangement
Na	2, 8, ①⟍	Na⁺	2, 8
	$\searrow$ 1e⁻		
Cl	2, 8, 7,	Cl⁻	2, 8, 8

◎ *Show how a lithium atom reacts with a fluorine atom. Name the compound formed.*

Bonding in elements and compounds

Noble gases have a stable arrangement of electrons and do not bond – they are monatomic. Metal atoms are arranged in a structure which allows the outer electrons to move through the structure. The atoms of all other elements are covalently bonded. Compounds formed between non-metal elements are **covalently bonded**. Compounds formed between metals and non-metals are **ionically bonded**.

Practice question

The grid shows the arrangement of atoms and molecules in pure substances and in mixtures.

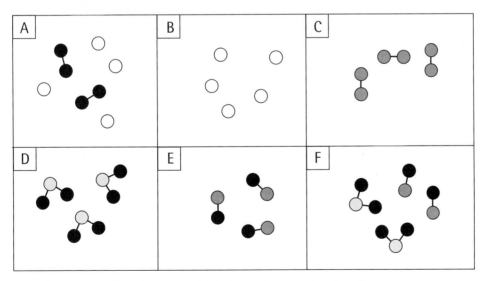

a) Identify the element which is made up of molecules.

b) Identify the mixture which is made up of compounds.

c) Identify the compound which could be hydrogen fluoride.

Properties of substances

⚙ Structure

Covalent substances exist as either molecules or networks.

The smallest molecules have two atoms – these are diatomic molecules.

Both elements and compounds can exist as diatomic molecules.

Covalent substances are mainly gases and liquids – some are solids with low melting and boiling points. Some covalent substances have very high melting and boiling points because they exist as networks. Networks are giant covalent structures (see diagrams, left).

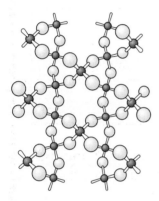

● C atoms

● O atoms ○ Si atoms

ⓒ Covalent substances which exist as networks need a lot of energy to break all the bonds and change state – this is why they are solids at room temperature. Covalent substances which exist as molecules have much lower melting points and boiling points because covalent bonds are not broken when they are melted or boiled. The molecules are held together by weak forces of attraction so little energy is needed to separate the molecules from each other when they melt or boil. The bigger the molecules the more energy is needed to separate them so the higher the melting and boiling points.

Ionic compounds all exist as solids at room temperature. The ions are arranged in a **crystal lattice** (see right).

They all have very high melting and boiling points.

ⓒ Ionic compounds have very high melting and boiling points because it takes a lot of energy to separate the ions from each other.

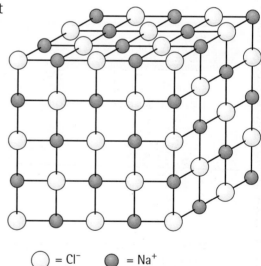

○ = Cl⁻ ● = Na⁺

❗ REMEMBER
You should know that the diatomic elements are: hydrogen $H_2(g)$, nitrogen $N_2(g)$, oxygen $O_2(g)$, fluorine $F_2(g)$, chlorine $Cl_2(g)$, bromine $Br_2(l)$ and iodine $I_2(s)$.

❗ REMEMBER
An electric current is a flow of charged particles. Electrons flow through metals. Ions flow through solutions and melts.

⚙ Electrical conductivity

All metals conduct electricity because they have outer electrons which can move through the structure. Graphite, which is a form of the element carbon, is the only non-metal conductor. Ionic substances do **not** conduct when solid because the ions are held so tightly they cannot move. When melted or dissolved in water, the ionic lattice breaks up and the ions are free to move, and so ionic melts and solutions conduct.

Covalent substances do not conduct in any state because they have no 'free' electrons and no ions.

Solubility

When an ionic substance dissolves in water the lattice breaks up completely.

Some covalent substances dissolve in water but many do not. Solid iodine doesn't dissolve well in water but does dissolve in alcohol. Nail varnish doesn't dissolve in water but does dissolve in propanone. Wax doesn't dissolve in water but does dissolve in hexane.

! **REMEMBER**
A solubility table is found on page 5 of the SQA Data Booklet.

Practice questions

1   Substances can be grouped as conductors or insulators.

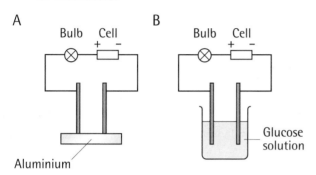

A   Bulb   Cell

B   Bulb   Cell

Aluminium

Glucose solution

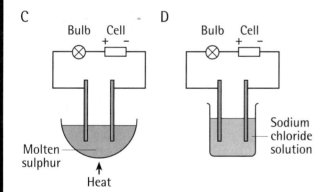

C   Bulb   Cell

D   Bulb   Cell

Molten sulphur

Heat

Sodium chloride solution

Identify the experiment(s) in which the bulb will light.

2   Jenny carried out electrical conductivity tests on several substances. The table below shows the results.

Substance	Conductor	Non-conductor
zinc	✓	
sodium chloride powder		✓
dilute nitric acid		
sodium chloride solution	✓	

a) Complete the table for dilute nitric acid.
b) Which substance conducts by a flow of electrons?
c) Why does sodium chloride powder not conduct electricity?

C 3 The table gives information about some substances.

Substance	Melting point/°C	Boiling point/°C	Conducts as: a solid	a liquid
A	1890	3380	yes	yes
B	963	1560	no	yes
C	1455	2730	yes	yes
D	-183	-164	no	no
E	1700	2230	no	no
F	712	1418	no	yes

a) Identify the two ionic compounds.
b) Identify the covalent molecular compound.
c) Identify the substance with a covalent network.

Electrolysis

 Passing a direct current (d.c.) through an ionic solution or melt causes a chemical reaction to occur at the electrodes. A direct current has a positive side and a negative side. An electrode, usually a carbon rod, is connected to each side. Positive ions are attracted to the negative electrode and negative ions are attracted to the positive electrode.

Metal ions are **positive** so the metal is formed at the **negative** electrode, and non-metal ions are **negative** so the non-metal element is formed at the **positive** electrode.

Examples:

(1) Electrolysis of copper(II) chloride solution.

Cu^{2+}(aq) ions are attracted to the negative electrode and copper metal is formed. Cl^-(aq) ions are attracted to the positive electrode and chlorine gas is formed.

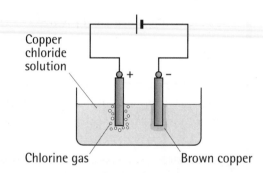

Copper chloride solution

Chlorine gas   Brown copper

> **REMEMBER**
> **Electrolysis** is the breaking up of an ionic compound when electricity is passed through it. An **electrolyte** is a substance which conducts due to the movement of ions. An **electrode** is usually a carbon or metal rod attached to the positive and negative sides of a d.c. power supply.

Ion-electron equations can be written to show what happens at each electrode. At the negative electrode, the Cu^{2+}(aq) ions are attracted and gain two electrons each:
$$Cu^{2+}(aq) + 2e^- \rightarrow Cu(s)$$

At the positive electrode, the Cl^-(aq) ions are attracted and lose an electron:
$$2Cl^-(aq) \rightarrow Cl_2(g) + 2e^-$$

(2) Electrolysis of molten lead(II) bromide.

Pb^{2+}(l) ions are attracted to the negative electrode and lead metal is formed. Br^-(l) ions are attracted to the positive electrode and liquid bromine is formed.

The ion-electron equations are as follows:

At the negative electrode:
$$Pb^{2+}(l) + 2e^- \rightarrow Pb(l)$$

At the positive electrode:
$$2Br^-(l) \rightarrow Br_2(l) + 2e^-$$

Write ion-electron equations to show what happens at each electrode when lithium iodide solution is electrolysed.

Name the products formed at each electrode when lithium iodide solution is electrolysed.

> **REMEMBER**
> You can find ion-electron equations on page 7 of the SQA Data Booklet.

Only ionic compounds can be electrolysed because there must be ions free to move through the solution or melt. The products of electrolysis can only be identified when a d.c. supply is used because there is a positive side and negative side to attract the oppositely charged ions.

⊚ Coloured ions

Some ionic compounds are coloured and this is related to the colour of the ions present, e.g. sodium chloride is colourless in solution so the sodium and chloride ions must be colourless. Sodium dichromate is orange and since the sodium ion is colourless, the dichromate ion must be orange.

The charge on an ion can be worked out by electrolysing a coloured compound, e.g. copper(II) dichromate.

The blue colour seen moving towards the negative electrode must be due to a positive ion – the $Cu^{2+}(aq)$ ion.

The orange colour seen moving towards the positive electrode must be due to a negative ion – the dichromate ion $[Cr_2O_7^{2-}(aq)]$.

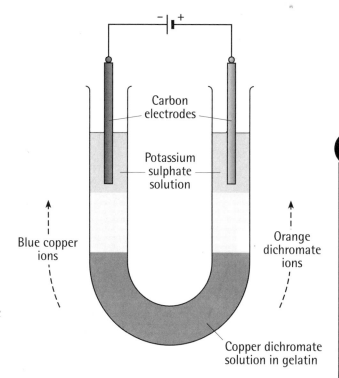

Practice questions

1   Paul connected a low voltage supply to carbon electrodes in a copper chloride solution. He wrote some notes in his jotter.

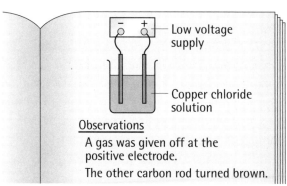

Observations
A gas was given off at the positive electrode.
The other carbon rod turned brown.

a) What type of experiment did Paul carry out?

b) Name the gas given off at the positive electrode.

c) Explain fully why the negative electrode "turned brown".

🄖 d) Write ion–electron equations for the reactions happening at each electrode.

2   The apparatus shown below was used to electrolyse molten lithium chloride.

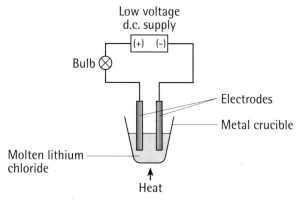

a) State what is meant by electrolysis.

b) Why does lithium chloride not conduct electricity when solid?

c) After the heat was removed, the lithium chloride changed into a solid but the bulb did not go out. Suggest a reason for this.

🄖 d) Write ion–electron equations for the reactions happening at each electrode.

Fuels

(SQA Topics: 5 &6)

📺 🔊 This section is about:

- fossil fuels

- fractional distillation

- pollution

- hydrocarbons.

Energy is needed to power vehicles, to make electricity in power stations and to keep our homes and schools warm. Most of our energy comes from coal, oil and gas – our fossil fuels. Fossil fuels were formed millions of years ago from dead plants and animals. They are vitally important to us but won't last forever – and they won't be replaced either!

The oil which comes out of the ground is known as crude oil. It is not pure enough to use as it is because it is a huge mixture of compounds. It has to be separated into smaller, more useful mixtures known as fractions, by a process called fractional distillation.

Unfortunately, the burning of fossil fuels produces gases which pollute the air – sulphur dioxide and oxides of nitrogen. Carbon dioxide is also produced and there are concerns that this is contributing to the greenhouse effect. Some scientists believe that we have to reduce the amounts of these gases released or our atmosphere will be polluted permanently.

Crude oil is a mixture of hydrocarbons, mainly alkanes. Alkanes react with oxygen from the air to produce lots of energy and so are good fuels. However, because of the way they are bonded, alkanes undergo few other reactions. Alkenes are another family of hydrocarbons but their structure differs from alkanes in a way that makes them more reactive. The alkenes are produced by cracking large, less useful alkanes.

The alkanes and alkenes are both made up of molecules in which the carbon atoms are joined to each other in chains. The cycloalkanes have the carbon atoms joined in rings. The alkanes, alkenes and cycloalkanes are all examples of a homologous series. A homologous series is a family of compounds which can be represented by a general formula and have similar chemical properties. Some compounds have the same molecular formula but different structures. This is known as isomerism.

g When you have finished this section, you should know the following at General Level:

■ A fuel is a substance which gives out energy when burned. Burning is the reaction of a substance with oxygen and is also known as combustion. An exothermic reaction is one which gives out heat.

■ The main gases in the air are oxygen and nitrogen, in the ration 1:4 and oxygen relights a glowing splint.

■ Oil, coal and gas are fossil fuels which took millions of years to form from dead animals and plants. They are finite resources.

■ Crude oil is a mixture of hydrocarbons. Hydrocarbons are compounds containing hydrogen and carbon only. Crude oil can be separated into small mixtures called fractions by a process called fractional distillation. Different fractions have different boiling ranges and differ in flammability and viscosity. The various fractions are mainly used as fuels.

■ When hydrocarbons burn completely they produce only carbon dioxide and water. Carbon dioxide turns lime water cloudy and water boils at 100°C and freezes at 0°C. Carbon monoxide is formed when there is insufficient air for complete combustion.

■ Sulphur dioxide and oxides of nitrogen are poisonous gases produced when some fossil fuels are burned and they pollute the air. Lead used to be added to petrol but is no longer added because it is poisonous.

■ Catalytic converters can remove pollutant gases and lean burn engines reduce the amounts produced.

■ The alkanes and alkenes are hydrocarbon families. You should know the names of the first eight alkanes and be able to write their molecular and structural formulae. The alkanes have the general formula C_nH_{2n+2} and are saturated.

■ You should know the names of the first six alkenes and be able to write their molecular and structural formulae. The alkenes have the general formula C_nH_{2n} and are unsaturated.

■ The bromine test can be used to tell a saturated compound from an unsaturated compound. Alkenes undergo addition reactions with bromine and hydrogen.

■ Alkenes are made by cracking long-chain alkanes.

C In addition, you should know the following at Credit Level:

■ The chain lengths of the molecules in each fraction and relate this to their use.

■ Bigger molecules have higher boiling points because it takes more energy to separate them. Smaller molecules are more flammable because they form gases easily. Compounds with bigger molecules are more viscous.

■ The transition metals in catalytic converters change harmful gases into harmless ones. Decreasing the fuel to air ratio improves the efficiency of combustion thus decreasing pollution.

■ You should be able to name the first four cycloalkanes.

■ A homologous series is a family of compounds which have a general formula and have similar chemical properties. The alkanes, alkenes and cycloalkanes are examples of homologous series.

■ Isomers are molecules with the same molecular formula but different structures. You should know how to draw isomers for alkanes, alkenes and cycloalkanes.

■ Catalysts allow cracking to happen at lower temperatures. A mixture of saturated and unsaturated molecules is produced.

Fuels

Fossil fuels

A fuel is a chemical which burns, giving out **heat energy**. Burning is also known as combustion and occurs when a fuel reacts with oxygen. Any reaction which gives out energy is called **exothermic**.

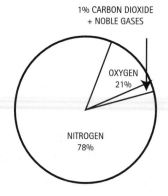

1% CARBON DIOXIDE + NOBLE GASES

OXYGEN 21%

NITROGEN 78%

REMEMBER
Oxygen relights a glowing splint. This is used as a test for oxygen.

REMEMBER
Oxygen makes up about one-fifth of the air.

Our main fuels are coal, oil and natural gas. These are the fossil fuels. Natural gas is mainly methane, a **hydrocarbon**, which means it is made of carbon and hydrogen only. It burns to produce carbon dioxide and water.

methane + oxygen → carbon dioxide + water

$$CH_4(g) + O_2(g) \rightarrow CO_2(g) + H_2O(l)$$

$$CH_4(g) + 2O_2(g) \rightarrow CO_2(g) + 2H_2O(l)$$

If carbon dioxide and water are produced when a fuel burns it proves the fuel must contain carbon and hydrogen as there is no carbon or hydrogen in the air!

Formation of fossil fuels

Fossil fuels have taken millions of years to form from what were once living things. This is where the name fossil fuel comes from. Fossil fuels are being used up very quickly and, because they take so long to form, they cannot be replaced. They are finite resources – they won't last forever. If we keep using them so intensively, they will eventually run out, causing a fuel crisis.

REMEMBER
Carbon dioxide turns lime water cloudy. Water boils at 100°C, and freezes at 0°C.

Coal

1. Coal started off as trees and other plants millions of years ago.	2. The plants died and were covered by mud and sand. More plants died and they too were covered.	3. Over millions of years a combination of pressure, heat and bacteria converted the plants into coal.

Oil

Oil and natural gas were thought to have formed in a similar way to coal except the starting material was the remains of dead sea creatures.

1. Millions of years ago sea creatures died and sank to the ocean bed.

2. Layers of mud and sand covered their remains.

3. Over millions of years, a combination of pressure, heat and bacteria converted the these remains into oil and gas.

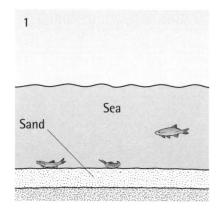

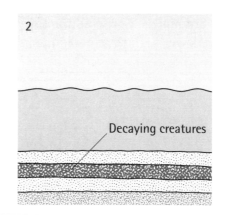

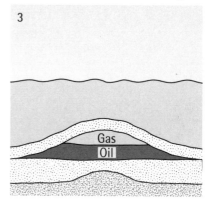

Practice questions

1 a) Describe the formation of coal.

   b) Name another fossil fuel.

2 a) What word is used to describe a reaction which produces heat?

   b) Name the products when propane reacts completely with oxygen.

3 Candle wax is a mixture of hydrocarbons. The apparatus shown below can be used to identify the products formed when a candle burns.

   a) Name the gas in air which is used up when a candle burns.

   b) Name two products formed when a candle burns.

   c) Why is test tube A cooled?

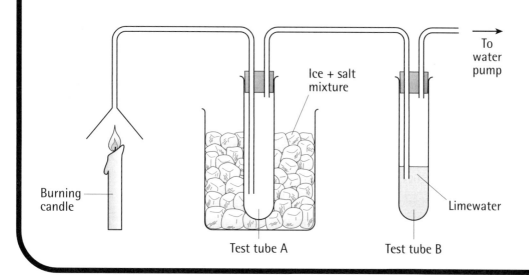

Fuels

Fractional distillation

📺 💿 Crude oil is a mixture of hydrocarbons and cannot be used for anything in this form. It has to be separated into smaller mixtures, known as **fractions,** which are collected over small temperature ranges. The process takes place in a fractionating tower. The tower has a high temperature at the bottom where hot oil is pumped in and the temperature decreases towards the top. As the hot molecules rise up the tower, they **condense** when they hit a temperature lower than their boiling points.

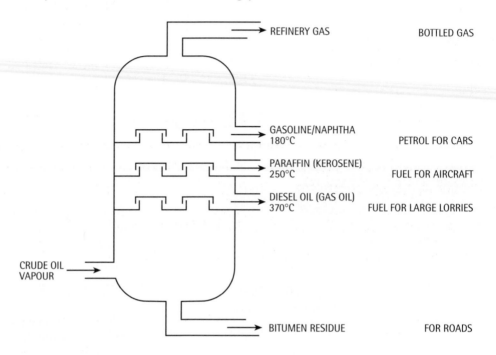

REFINERY GAS — BOTTLED GAS

GASOLINE/NAPHTHA 180°C — PETROL FOR CARS

PARAFFIN (KEROSENE) 250°C — FUEL FOR AIRCRAFT

DIESEL OIL (GAS OIL) 370°C — FUEL FOR LARGE LORRIES

CRUDE OIL VAPOUR

BITUMEN RESIDUE — FOR ROADS

Residue

❶ REMEMBER A substance that is viscous is very syrupy (thick). A substance that is flammable catches fire very easily.

Fractions which contain large molecules have high boiling points and condense at the bottom of the tower. Bitumen is part of this fraction and is used in road making. It is very viscous because the molecules are so long. They don't evaporate easily and they are not very flammable. This means bitumen is not a good fuel.

Diesel

The diesel fraction has smaller molecules than residue and condenses further up the tower where the temperature is lower. Diesel is less viscous than bitumen and is flammable – this makes it useful as a fuel.

Gasoline/Naphtha

This fraction has even smaller molecules and condenses nearer the top of the tower. The gasoline fraction is in great demand because it is a source of petrol – one of our most important fuels.

C   At Credit Level, you should know the typical ranges of molecular length and be able to relate molecular length to the use of each fraction. This is summarised in the table below.

Name of fraction	Boiling range/°C	Carbons per molecule	% crude oil	Use
refinery gas	<20	C1– C4	1-2	gaseous fuels and blending with petrol; feedstock for other chemicals
gasoline/ naphtha	20-180	C5-C10	15-30	petrol for cars (gasoline); production of other chemicals (naphtha)
kerosene	160-250	C10-C16	10-15	jet fuel; heating fuel
diesel oil	220-370	C13-C25	15-20	diesel fuel; heating fuel
residue	>370	>C25	40-50	fuel oil; lubricating oils; bitumen and asphalt for roads and roofing

Practice questions

1   The table gives information about fractions obtained from crude oil.

Fraction	Number of carbon atoms per molecule	Boiling point range/°C
A	1–4	under 20
B	4–12	20–175
C	9–16	150–240
D	15–25	220–250
E	20–70	250–350

a) Identify the two fractions which contain butane.

b) Identify the liquid fraction with the lowest boiling point range.

2   Crude oil arriving at the BP refinery in Grangemouth is separated into different fractions. Use the diagram on page 32 to answer these questions.

a) Which liquid fraction is most flammable?

b) Give a use for the residue.

C c) In which fraction will pentane be found? You may wish to use page 6 of the SQA Data Booklet to help you.

C d) Why is the gas oils fraction more viscous than the kerosene fraction?

C e) Fractions which are surplus to requirements can be cracked. Give a reason for cracking fractions.

Pollution

Extracting, transporting and using fuels can cause pollution. Digging coal out of the ground can leave the countryside scarred forever so the area has to be reinstated. Crude oil is transported all over the world in giant tankers. Oil spills pollute water and beaches and this, in turn, kills fish and sea birds.

Burning fuels produces carbon dioxide and water. There are, however, other elements in fuels. Coal, for example, contains **sulphur**. When it is burned, poisonous sulphur dioxide is released into the atmosphere.

$$\text{sulphur} + \text{oxygen} \rightarrow \text{sulphur dioxide}$$
$$S(s) \quad + \quad O_2(g) \quad \rightarrow \quad SO_2(g)$$

Sulphur dioxide dissolves in water in the atmosphere and forms an acid solution which contributes to acid rain.

Low sulphur coal can be burned to reduce pollution. Sulphur dioxide can be removed in the chimneys of power stations before it reaches the atmosphere.

Lead used to be added to petrol to make it burn more efficiently inside the engine. The poisonous lead was then released into the atmosphere through the exhaust. Unleaded and lead replacement petrol are now used to avoid this problem. Inside a car engine there is a limited supply of oxygen and incomplete combustion of petrol takes place. This means that **unburned carbon, hydrocarbons** and **carbon monoxide** (a poisonous gas) are released into the atmosphere. In a petrol engine, a petrol and air mixture is 'sparked' to ignite it. This produces oxides of nitrogen (NO_X) which are poisonous and contribute to acid rain.

REMEMBER These are the main pollutants from burning fossil fuels: carbon monoxide (CO), sulphur dioxide (SO_2) and oxides of nitrogen (NO_X).

Pollutant gases: sources and effects

Pollutant	Main source	Effect
Carbon monoxide	Vehicle engines and industry	Poisonous
Sulphur dioxide	Burning fossil fuels in power stations	Forms acid rain
Hydrocarbons	Burning fuels in vehicles and factories	Irritating, toxic compounds formed in air
Oxides of nitrogen	Vehicle engines	Forms acid rain

Emissions of carbon monoxide and oxides of nitrogen are reduced in modern cars because they have **catalytic converters** attached to their exhausts. Lean burn engines use a higher air/fuel ratio so less carbon monoxide is formed.

G Catalytic converters use transition metals like platinum and rhodium to act as catalysts. The polluting gases are converted into harmless gases:

$$CO, NO_x \rightarrow \text{catalytic converter} \rightarrow CO_2, N_2$$

In a lean burn engine, because there is more air, more carbon dioxide is produced instead of carbon monoxide.

Practice questions

1. Natural gas is mainly methane. It is burned in gas fires. Each year about fifty people are killed by a poisonous gas produced by faulty fires.

   a) Name the poisonous gas.

   b) Explain why this gas is produced.

2. Oil spilt at sea can cause damage to the environment. This table shows the sources of oil pollution of the sea.

Source of oil pollution	%
Industrial waste	60
Oil industry	14
Natural sources	10
Shipping	16

   Draw a bar chart to show this information.

3. Many cars are fitted with catalytic converters. They change harmful gases produced in the engine into harmless gases (see diagram below).

   a) Oxides of nitrogen react with carbon monoxide in the converter. Name the two harmless gases produced.

   b) Name a metal which is used as a catalyst in a catalytic converter.

   c) Suggest why the catalyst is spread over a large surface area.

   d) State another way of reducing pollution from a petrol engine.

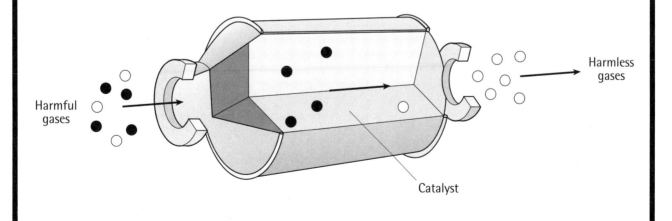

Harmful gases

Harmless gases

Catalyst

Fuels

Hydrocarbons

There are three hydrocarbon families to study: alkanes, alkenes and (at Credit level) cycloalkanes.

The alkanes

Their names end in **-ane**. You have to know the names of the first eight: they are methane, ethane, propane, butane, pentane, hexane, heptane and octane. They have the general formula C_nH_{2n+2}, where n = 1, 2, 3, etc. They have only C–C single bonds which means they are saturated. Most are good fuels.

36

REMEMBER In the exam, remember that alkanes are saturated, and that alkenes are unsaturated and undergo addition reactions.

Name	Molecular formula	Full structural formula	Shortened structural formula
Methane	CH_4	H | H—C—H | H	CH_4
Ethane	C_2H_6	H H | | H—C—C—H | | H H	CH_3CH_3

The alkenes

Their names end in **-ene**. The first part of the name is the same as the alkanes. You have to know the names of the first five: they are ethene, propene, butene, pentene and hexene. They have the general formula C_nH_{2n}, where n = 2, 3, 4 etc. They have one C=C double bond which means they are unsaturated. The double C=C bond explains why the first alkene is ethene (C_2H_4).

Name	Molecular formula	Full structural formula	Shortened structural formula
Ethene	C_2H_4	H H \ / C=C / \ H H	CH_2CH_2

The carbon atoms in alkanes and alkenes are arranged in open chains.

Alkanes decolourise brown bromine solution slowly. This can be used as a test to tell an unsaturated compound (alkene) from a saturated compound (alkane). The reaction of an alkene with bromine and hydrogen is an example of an addition reaction.

ethene (alkene) + hydrogen $\rightarrow$ ethane (alkane)

Alkenes are made up by breaking up long chain alkanes obtained from fractional distillation of crude oil. This is called **catalytic cracking**.

The use of a catalyst allows cracking to happen at a lower temperature.

A mixture of alkanes and alkenes is obtained because there are not enough hydrogens for all the smaller molecules formed to be saturated.

$$C_{10}H_{22} \quad \rightarrow \quad C_8H_{18} \quad + \quad C_2H_4$$
$$\text{alkane} \qquad\qquad \text{alkane} \qquad \text{alkene}$$

C The cycloalkanes

The carbon atoms in the cycloalkanes are arranged in rings. They have the general formula C_nH_{2n}, where n = 3, 4, 5 etc. The smallest number of atoms which can form a ring is three.

The first part of their name is always **cyclo-**. The second part of their name comes from the corresponding alkane, e.g. cyclopropane. You have to be able to name the following cycloalkanes: cyclopropane, cyclobutane, cyclopentane, cyclohexane.

Name	Full structural formula	Molecular formula
Cyclopropane	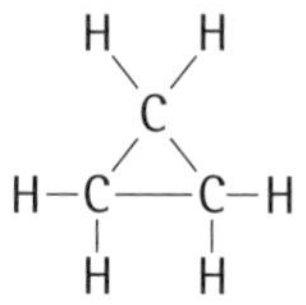	C_3H_6
Cyclobutane	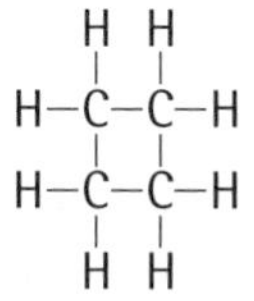	C_4H_8

C Homologous series

Families of compounds with similar chemical properties which can be represented by a general formula are called **homologous series**. Alkanes, alkenes and cycloalkanes are homologous series. The steady change in boiling point in any homologous series is due to the gradual increase in the size of the molecules. The bigger they get, the stronger the forces between the molecules and the more energy it takes to separate them.

C Isomerism

A compound which has more than one possible structural formula is said to show **isomerism**.

Butane (C_4H_{10}) has two isomers:

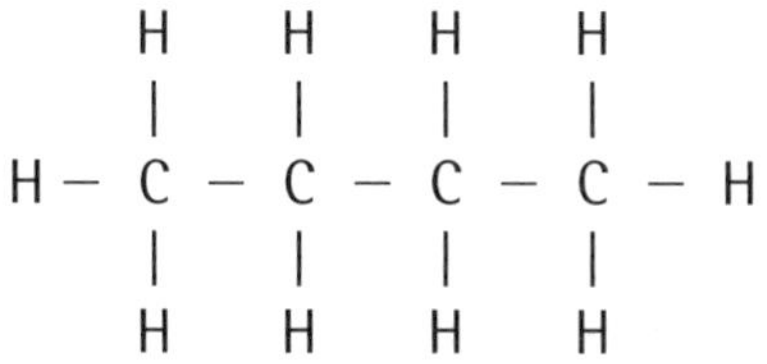

Straight chain

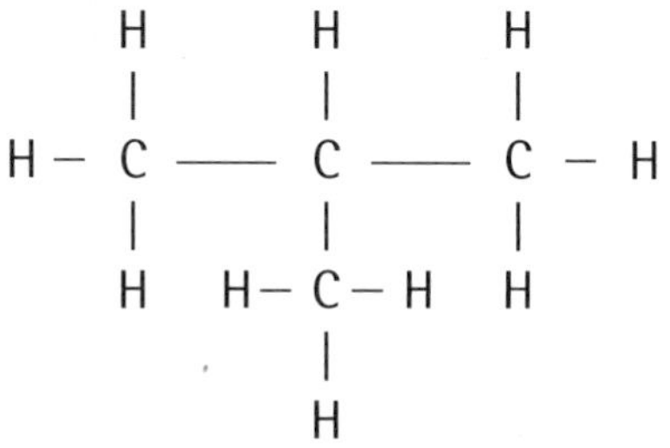

Branched chain

> **REMEMBER**
> Isomers have the same elements and the same number of atoms, but a different structure.

Fuels

Acids and alkalis

📺 ◉ This section is about:

- acids and alkalis

- reaction of acids.

Acids and alkalis play an important part in our lives. We find them in our home and they occur in our bodies. They are important industrially for making other chemicals. Non-metal oxides which dissolve in water form acids. Soluble metal oxides and hydroxides form alkalis. Acids produce $H^+(aq)$ ions in solution. Alkalis produce $OH^-(aq)$ ions in solution.

Acids turn pH indicator shades of red and have a pH number less than 7. Alkalis turn pH indicator shades of blue and have a pH number greater than 7. Substances with a pH equal to 7 are neutral. When acids are diluted their pH increases towards 7. When alkalis are diluted their pH drops towards 7.

Alkalis and metal carbonates react with acids and the pH of the acid rises towards 7. The hydrogen ions react to form water. This is called neutralisation. Substances which take part in neutralisation reactions with acids are generally called bases. Neutralisers are not just used in the laboratory – they are part of our everyday life. Acid indigestion and the pain caused by acid stings can be relieved by neutralisers. Farmers reduce the acidity of soil by adding lime, a neutraliser.

Soluble and insoluble salts can be made in the laboratory. Soluble salts are made by reacting in acid with a base. The salt formed is recovered from solution by evaporation. Insoluble salts are prepared by mixing solutions containing the ions required in the salt. The insoluble salt forms a precipitate which is washed and filtered.

Sulphur dioxide and oxides of nitrogen are released into the atmosphere when fossil fuels burn. These gases dissolve in water in the atmosphere, making it acidic. This is called acid rain. Acid rain attacks unprotected metals like iron and stone containing carbonate. This can weaken structures and be extremely dangerous.

g When you have finished this section, you should know the following at General Level:

■ Acids and alkalis are all around us. You should be able to give an example of an acid and alkali found in the laboratory and at home.

■ Non-metal oxides which dissolve in water produce acid solutions, whereas soluble metal oxides or hydroxides produce alkaline solutions.

■ pH is a continuously numbered scale which ranges from 1–14. Acids have a pH less than 7 and alkalis have a pH greater than 7. Water and neutral solutions have a pH= 7. pH paper or universal indicators can be used to test the pH of a solution.

■ Ions are present in acids, alkalis and water. The concentration of ions in water is small. An acidic solution contains $H^+(aq)$ ions, an alkaline solution contains $OH^-(aq)$ ions.

■ When an acid is diluted, it becomes less acidic and its pH increases towards 7, and when an alkali is diluted, it become less alkaline and its pH decreases towards 7.

■ When an acid reacts, its pH increases towards 7. Acids react with alkalis to form water plus a salt.

■ When an acid reacts with an alkali the pH of the alkali moves towards 7. Acids react with carbonates to form water, carbon dioxide and a salt.

■ Sulphuric acid forms sulphates, nitric acid forms nitrates and hydrochloric acid forms chlorides.

■ Reactions of acids which produce water are called neutralisation reactions. $H^+(aq)$ ions react to form water during neutralisation.

■ Adding lime to fields and taking indigestion tablets are everyday examples of neutralisation.

■ Reactive metals react with acids to produce hydrogen gas and a salt. $H^+(aq)$ react to form $H_2(g)$ when a metal reacts with an acid. Hydrogen gas burns with a 'pop' – this is the test for hydrogen.

■ Precipitation is the reaction of two solutions to form an insoluble precipitate. You should know how to use the SQA Data Booklet to name a precipitate.

■ Sulphur dioxide and oxides of nitrogen dissolve in water present in the atmosphere to form acid rain. Iron structures and stone containing carbonate can be damaged because they react with acid rain.

C In addition, you should know the following at Credit Level:

■ In water and neutral solutions the concentration of $H^+(aq)$ and $OH^-(aq)$ is the same.

■ As an acid is diluted the concentration of $H^+(aq)$ decreases, as an alkali is diluted the concentration of $OH^-(aq)$ decreases.

■ A base is a substance which undergoes a neutralisation reaction with an acid. Soluble bases dissolve in water to form alkalis.

■ Insoluble metal carbonates or metal oxides are used to prepare salts.

■ When an acid is neutralised by an alkali, $H^+(aq)$ reacts with $OH^-(aq)$ to form water. When an acid is neutralised by a carbonate, $H^+(aq)$ reacts with $CO_3^{2-}(aq)$ to form water and carbon dioxide ($H_2O + CO_2$).

Acids and alkalis

Acids and alkalis

Acids

REMEMBER
Memorise names and formulae of these laboratory acids.

We come in contact with acids all the time. We eat them, we drink them – we even have acid in our stomach. Jams, sauces, fruits, fizzy drinks, lemon juice and vinegar are all acidic.

Acids found on the shelf in the laboratory include:
hydrochloric acid – HCl(aq), sulphuric acid – H_2SO_4(aq), nitric acid – HNO_3(aq)

Acids are formed when non-metal oxides dissolve in water

e.g. sulphur dioxide + water $\rightarrow$ sulphurous acid
$$SO_2(g) + H_2O(l) \rightarrow H_2SO_3(aq)$$

Write a word formula equation for carbon dioxide dissolving in water to form carbonic acid.

Alkalis

REMEMBER
The (aq) symbol means the acids and alkalis are in solution.

Alkalis are all around us, in the home and in the laboratory. Sodium bicarbonate is used in cooking and to help relieve acid indigestion. Milk of magnesia also relieves indigestion. Ammonia can be used as an oven cleaner.

Alkalis found on the shelf in the laboratory include:
sodium hydroxide – NaOH(aq), potassium hydroxide – KOH(aq), ammonia – NH_3(aq)

REMEMBER
You can find out if an oxide or hydroxide is soluble by checking the solubility table on page 5 of the SQA Data Booklet.

Alkalis are made when a soluble metal oxide or hydroxide dissolves in water.

e.g. sodium oxide + water $\rightarrow$ sodium hydroxide
$$Na_2O(s) + H_2O(l) \rightarrow NaOH(aq)$$
$$Na_2O(s) + 2H_2O(l) \rightarrow 2NaOH(aq)$$

Write the word, formula and balanced equation for potassium oxide (K_2O) dissolving in water.

pH

We can tell if a solution is acid or alkaline by testing its pH using Universal indicator or pH paper. pH is a continuously numbered scale which ranges from 1 to 14. Acids have a pH **less than 7** which shows up red or shades of red with pH indicator. Alkalis have a pH **greater than 7** which shows up blue or shades of blue with pH indicator. Substances like water which are neither acidic nor alkaline have a pH of 7 and are said to be neutral. This shows as light green with pH indicator.

REMEMBER
Acid: pH < 7; alkali: pH > 7; neutral: pH = 7.

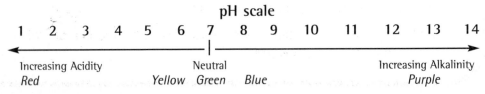

pH scale

| 1 | 2 | 3 | 4 | 5 | 6 | 7 | 8 | 9 | 10 | 11 | 12 | 13 | 14 |

Increasing Acidity | Neutral | Increasing Alkalinity
Red | *Yellow Green Blue* | *Purple*

Ions in solution

All acids conduct electricity showing that ions are present. All acids produce hydrogen gas at the negative electrode during electrolysis indicating that acids contain the hydrogen ion, $H^+(aq)$ in solution.

All alkalis conduct electricity showing that ions are present. Alkalis contain the hydroxide ion, $OH^-(aq)$. Water is a poor conductor of electricity because it has a low concentration of ions.

C Water and neutral solutions have $H^+(aq)$ ions and $OH^-(aq)$ ions, but in equal concentrations. They are neither acidic nor alkaline because the ions counteract the effect of each other.

C When an acidic substance is dissolved in water, the number of $H^+(aq)$ ions in solution **increases**. Acidic solutions have more $H^+(aq)$ ions than pure water.

C When an alkaline substance is dissolved in water, the number of $OH^-(aq)$ ions in solution **increases**. Alkaline solutions have more $OH^-(aq)$ ions than pure water.

📺 ⏵ Dilution

As an acid is diluted with water, the solution becomes less acidic and so the pH rises towards 7. As an alkali is diluted with water, the solution becomes less alkaline and so the pH falls towards 7.

C pH is a measure of $H^+(aq)$ concentration. As an acid is diluted, the $H^+(aq)$ ions spread out and the solution becomes less concentrated. So the pH rises towards 7. As an alkali is diluted, the $OH^-(aq)$ ions spread out and the solution becomes less concentrated so the pH falls towards 7.

Practice question

The chart shows the pH of some common substances.

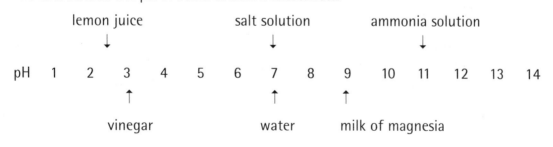

a) Identify the two substances which are acids.

b) Identify the two substances which will show a decrease in pH when they are diluted with water.

Reactions of acids

📺 ⏱ In this section, you have to know the products formed when acids react with alkalis, metal carbonates and metals. You also have to be able to write word and formula equations for the various reactions.

When acids react, their acidity is reduced because the hydrogen ions are removed during the reaction. This means the pH of the acid moves up towards 7. **Salts** are formed when acids react – you should remember the name endings for the salts formed from different acids.

Acid	Salt
hydrochloric	chloride
sulphuric	sulphate
nitric	nitrate

Acids and alkalis

❗ REMEMBER Take frequent short breaks when revising.

When acids react with alkalis, water and a salt are produced. The pH of the acid moves up towards 7 and the pH of the alkali moves down towards 7, i.e. neutral.

acid + alkali   → water + salt

hydrochloric acid + sodium hydroxide → water + sodium chloride

$HCl(aq)$ + $NaOH(aq)$ → $H_2O(l)$ + $NaCl(aq)$

◎ *Write word and formula equations for:*
  i. sulphuric acid reacting with sodium hydroxide
  ii. nitric acid reacting with potassium hydroxide.

Acids and metal carbonates

When acids react with metal carbonates water, carbon dioxide and a salt are formed. The pH of the acid moves up towards 7.

acid   + carbonate → water + carbon dioxide + a salt

sulphuric + calcium → water + carbon dioxide + calcium sulphate
 acid   carbonate

$H_2SO_4(aq) + CaCO_3(s)$ → $H_2O(l)$ + $CO_2(g)$ + $CaSO_4(aq)$

◎ *Write word and formula equation for:*
  i. hydrochloric acid reacting with sodium carbonate (Na_2CO_3)
  ii. nitric acid reacting with potassium carbonate (K_2CO_3).

Any reaction of an acid which results in water being produced is called **neutralisation**. In every neutralisation reaction hydrogen ions are removed from the solution as water. Everyday examples of neutralisation are:

- the reduction of soil acidity by adding lime
- the addition of lime to lochs which have become too acidic
- the treatment of acid indigestion.

REMEMBER
A neutral solution will contain equal numbers of H^+ and OH^- ions.

Ⓒ Bases

A substance which undergoes a neutralisation reaction with an acid is called a base. Soluble bases form alkaline solutions. Copper(II) oxide is a base because it neutralises an acid but does not form an alkaline solution because it is insoluble. Potassium hydroxide is a base and forms an alkaline solution because it is soluble.

◎ *Use the solubility table to select two soluble and two insoluble bases.*

Ⓒ Insoluble metal carbonates and metal oxides are often used to prepare salts by the following process:

(1) The solid base is added to the acid until no more reacts.
(2) The excess base is then filtered off leaving the required salt in solution.
(3) The solid salt can be obtained by evaporating off the water from the solution.

REMEMBER
A table showing the solubilities of bases (oxides, hydroxides and carbonates) is found on page 5 of the SQA Data Booklet.

Ⓒ Ions reacting

In any neutralisation reaction it is the $H^+(aq)$ ions which react to form water.

acid + alkali

$$H^+(aq) + OH^-(aq) \rightarrow H_2O(l)$$

acid + carbonate

$$H^+(aq) + CO_3^{2-}(aq) \rightarrow H_2O(l) + CO_2(g)$$

Any ions which don't react are called **spectator ions**.

Acid + metal

Some metals react with acids but hydrogen gas, rather than water, is produced so this is not a neutralisation. A salt is produced.

acid	+	reactive metal	→ hydrogen +	a salt
hydrochloric acid +		zinc	→ hydrogen +	zinc chloride
sulphuric acid +		magnesium	→ hydrogen +	magnesium sulphate

In each case, hydrogen ions gain electrons and hydrogen molecules are formed.

$$2H^+(aq) + 2e^- \rightarrow H_2(g)$$

ions   molecules

◎ *Name the products formed when hydrochloric acid reacts with magnesium.*

Precipitation

Precipitation is the reaction of two solutions to form an insoluble product called a precipitate. Reactions of acids form soluble salts. Insoluble salts are formed by precipitation.

The solutions of two soluble salts, containing the ions that will create an insoluble salt, are mixed. The mixture is filtered and the precipitate washed and dried.

The solubility table on page 5 of the SQA Data Booklet is used to select the soluble salts.

Example: barium sulphate

Any soluble barium compound could be used, e.g. **barium** nitrate.

Any soluble sulphate could be used, e.g. magnesium **sulphate**.

barium   +   magnesium →   barium   +   magnesium
nitrate(aq)   sulphate(aq)   sulphate(s)   nitrate(aq)

◉   *Select two solutions which would react to form lead(II) iodide and write a word equation for the reaction.*

Acid rain

The burning of fossil fuels produces acidic gases. Burning coal produces sulphur dioxide and burning petrol produces oxides of nitrogen. Both of these gases are oxides of non-metals and dissolve in water in the atmosphere producing an acid mixture known as **acid rain**.

Acid rain will react with unprotected metal structures, most commonly iron. Car bodywork and the Forth Rail Bridge are made from steel, which is mainly iron. The iron loses its strength and becomes weakened and potentially dangerous.

Acid rain will also react with **carbonate** in the stonework of some buildings. The stonework is eroded and can break off, weakening the structure. Many statues are made of marble, a carbonate, and show signs of wearing due to the effects of acid rain over many years.

Plants and animals also suffer from the effects of acid rain. Lochs become contaminated and this can kill fish and other animals in the water. Algae can also grow on the surface of the water and this reduces the amount of oxygen in the water. Heavy acid rainfall washes nutrients out of the soil so plants don't grow so well.

Practice questions

1 There are many different magnesium compounds.

A	B	C
magnesium bromide	magnesium carbonate	magnesium chloride

D	E	F
magnesium nitrate	magnesium oxide	magnesium sulphate

(a) Identify the two compounds which are insoluble.

(You may wish to use page 5 of your SQA Data Booklet to help you.)

(b) Identify the compound(s) which will neutralise an acid.

2 Katie carried out an experiment to make copper chloride crystals.

Workcard – Making Salts

Preparation of copper chloride crystals

1 Measure 50 cm³ of dilute hydrochloric acid into the beaker.
2 Add a spatula-full of copper carbonate to the beaker and stir.
3 Continue to add copper carbonate until no more reacts.
4 Filter.
5 Leave the copper chloride solution to crystallise.

a) Name the type of reaction that took place.

b) How did Katie know that no more copper carbonate was reacting?

c) Why was the solution filtered?

d) Why can copper chloride not be made by adding copper to dilute hydrochloric acid?

C 3 The grid below shows pairs of chemicals.

A	B
$Mg(s) + HCl(aq)$	$NaOH(aq) + H_2SO_4(aq)$

C	D
$Cu(s) + H_2SO_4(aq)$	$Zn(s) + AgNO_3(aq)$

E	F
$CuSO_4(aq) + NaCO_3(aq)$	$CaCO_3(s) + HCl(aq)$

a) Identify the pair(s) which would react to produce water.

b) Identify the pair which would not react.

Making electricity

This section is about:

- batteries

- the electrochemical series

- reactions in cells.

46

We use batteries as a portable source of electricity. Batteries are also known as cells. They come in all shapes and sizes and supply various amounts of electricity. Chemical reactions take place inside batteries and this produces electricity. Electricity is a flow of electrons through a wire. Most commercial batteries contain metals and an electrolyte. An electrolyte is an ionic solution and it completes the circuit in a battery.

Once the chemicals in a battery are used up, the battery goes 'flat' and stops making electricity. These batteries have to be replaced. Some batteries like the lead–acid battery in a car are rechargeable and can be used over and over again. Batteries have a number of advantages over mains electricity, such as being portable but there are some disadvantages too, such as not being able to supply a high enough current for most household appliances.

When pairs of different metals are connected in a cell, a voltage is produced. Different metal pairings give rise to different voltages and the voltages can be used to put metals in order of how well they form ions. This is the electrochemical series. The electrochemical series can be used to explain why displacement reactions occur.

Electricity can also be produced by connecting different metals in solutions of their own ions. The metals are connected to each other by wires which allow electrons to flow from the metal higher in the electrochemical series to the one lower. The solutions are connected by an 'ion bridge' which allows ions to move from one solution to another. It is possible to produce electricity from cells which do not involve metals.

Reactions take place at the electrodes in a cell. The substance which loses electrons is said to be oxidised. The substance which gains electrons is said to be reduced. Oxidation and reduction always take place together. This type of reaction is called a redox reaction.

FactZONE

g When you have finished this section, you should know the following at General Level:

- A battery is also known as a cell. In a battery, electricity comes from a chemical reaction. Most batteries have to be replaced when the chemicals are used up in the reaction. Some batteries are rechargeable e.g. the lead–acid battery in a car.

- An electrolyte is an ionic solution which completes the circuit in a battery. Ammonium chloride in a battery is acting as an electrolyte.

- Both batteries and mains electricity have advantages and disadvantages:

  Batteries
  Advantage: portable
  Disadvantage: low current produced

  Mains
  Advantage: current high enough for heavy duty machines
  Disadvantage: high risk of shock

- A cell can be made up of two different metals connected together with an electrolyte. The voltage between different metals varies and this results in the electrochemical series which shows the metals in order of how well they form ions. (The electrochemical series is on page 7 of the SQA Data Booklet). A metal can displace ions of a metal lower in the electrochemical series from a solution containing ions of the lower metal.

- Cells can be set up with metals in solutions of their own ions. The ion bridge in a cell completes the circuit.

- Electrons always flow through the wires connecting metals in a cell. The direction of electron flow in a cell is from the metal higher in the electrochemical series to the one lower in the series.

c In addition, you should know the following at Credit Level:

- The position of a metal in the electrochemical series can be used to predict whether a displacement reaction will take place and what will be observed.

- The reaction of acids with metals leads to hydrogen being placed in the electrochemical series between lead and copper.

- The ion bridge in a cell allows ions to move between the solutions.

- Oxidation is loss of electrons by a reactant in any reaction and reduction is gain of electrons by a reactant in any reaction. Oxidation and reduction go on together in a redox reaction.

- Cells can be made with electrodes which are non-metals.

- Ion–electron equations are listed in the SQA Data Booklet. They are written as reductions but can be reversed to get the oxidation equation.

Batteries and cells

Chemical cells

A chemical cell converts **chemical** energy into **electrical** energy. When two or more cells are joined together it is called a **battery**. However, an individual cell is often called a battery so you should be familiar with both terms.

The zinc–carbon battery

The zinc-carbon battery is the most common type of battery which is used to power machines which don't require a lot of electricity, e.g. a personal stereo or torch.

When the battery is connected into an appliance, a chemical reaction inside the battery produces electrons which flow along a metal conductor and, in the case of a torch, through the bulb which then lights. The ammonium chloride acts as an **electrolyte**.

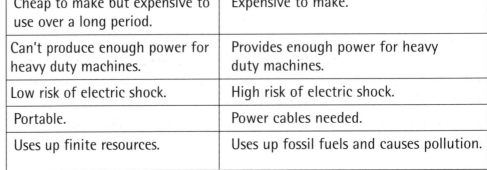

Positive electrode

Zinc case (negative electrode)

Ammonium chloride soaked paper

Manganese dioxide and carbon mixture

Carbon rod

An electrolyte is an ionic compound which is needed to complete the circuit. The electrolyte is normally in solution to allow the ions to move but, in the zinc-carbon battery, it is in the form of a paste to avoid it leaking out.

The chemicals in a battery eventually get used up and the battery goes 'flat'. Zinc-carbon batteries have to be replaced but some batteries are re-chargeable. The lead–acid battery in a car is rechargeable. The battery gets discharged while trying to start the engine but gets recharged once the engine is turning.

The table below shows advantages and disadvantages of batteries and mains electricity.

REMEMBER A single cell is often called a battery, but a battery is the correct term for two or more connected cells.

Batteries	Mains
Cheap to make but expensive to use over a long period.	Expensive to make.
Can't produce enough power for heavy duty machines.	Provides enough power for heavy duty machines.
Low risk of electric shock.	High risk of electric shock.
Portable.	Power cables needed.
Uses up finite resources.	Uses up fossil fuels and causes pollution.

The electrochemical series

Electricity can be produced in a cell made by connecting different metals together – an electrolyte must be used to complete the circuit.

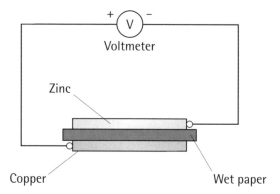

The voltage between different pairs of metals varies and, if one metal is kept the same, the voltages which all the other metals produce when connected into the cell can be compared. The metals which form ions most readily give the highest voltage. The results from this type of experiment has led to the **electrochemical series**, which can be found on page 7 of the SQA Data Booklet. The metal higher in the electrochemical series will give electrons to a metal lower in the series when the two are connected in a cell. The further apart the metals, the higher the voltage.

> **REMEMBER**
> The further apart the two metals are in the electrochemical series, the higher the voltage produced.

Using the electrochemical series

A metal can **displace** another metal from a solution if the metal being added is higher in the electrochemical series. Zinc metal for example, will displace copper ions from a solution of copper(II) sulphate because zinc is higher in the electrochemical series than copper.

$$Zn(s) + CuSO_4(aq) \rightarrow Cu(s) + ZnSO_4(aq)$$

C   The electrochemical series can be used to predict whether displacement reactions will occur and what will be observed. Hydrogen appears in the electrochemical series between lead and copper, even though it is not a metal. Experiments have shown that metals above copper can displace hydrogen ions from solution and metals below lead cannot.

More cells

Electricity can be produced in a cell by connecting two different metals in solutions of their metal ions. The solutions are usually in separate beakers and connected by an 'ion bridge', which is usually a piece of filter paper dipped in an electrolyte, in order to complete the circuit.

C   The ion bridge actually allows ions to move from one beaker to the other.

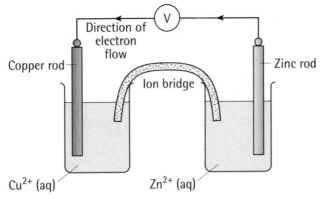

The diagram shows electron flowing from zinc to copper through the connecting wire. This is because zinc is higher in the electrochemical series than copper.

G The zinc loses electrons and forms ions.

The ion-electron equation is:

$$Zn(s) \rightarrow Zn^{2+}(aq) + 2e^-$$

Loss of electrons is known as **oxidation**.

The electrons from the zinc are gained by the copper ions and copper metal is formed.

The ion-electron equation is:

$$Cu^{2+}(aq) + 2e^- \rightarrow Cu(s)$$

Gain of electrons is known as **reduction**.

Reactions in which reduction and oxidation take place are called **redox** reactions. Reduction and oxidation always take place together.

G Cells without metals

Electricity can be produced in a cell where one or even both half cells do not involve a metal.

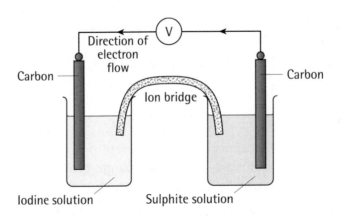

The direction of electron flow indicates that the sulphite ion, $SO_3^{2-}(aq)$, is losing electrons, i.e. is being oxidised. The ion–electron equation (from the SQA Data Booklet) is:

$$SO_3^{2-}(aq) + H_2O(l) \rightarrow SO_4^{2-}(aq) + 2H^+(aq) + 2e^-$$

The iodine, $I_2(aq)$, is gaining the electrons, i.e. is being reduced. The ion–electron equation is:

$$I_2(aq) + 2e^- \rightarrow 2I^-(aq)$$

Practice questions

1 This is a diagram of a dry cell commonly used in torches.

Zinc case (−) ——— Carbon rod (+)

——— Ammonium chloride paste

a) i) Why is the ammonium chloride used as a paste and not as a dry powder?

Eventually the cell stops producing electricity and has to be replaced.

  ii) Explain why the cell stops producing electricity.

b) Car batteries are made from much heavier wet cells containing lead plates and sulphuric acid. Suggest the main advantage of this lead–acid battery.

C 2 Ion–electron equations can be used to show the gain and loss of electrons in chemical reactions.

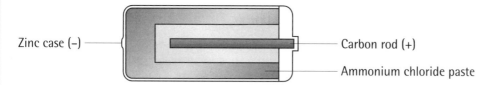

A		B	
$Fe(s) \rightarrow Fe^{2+}(aq) + 2e^-$		$Fe^{2+}(aq) + 2e^- \rightarrow Fe(s)$	
C		**D**	
$Fe^{2+}(aq) \rightarrow Fe^{3+}(aq) + e^-$		$Fe^{3+}(aq) + e^- \rightarrow Fe^{2+}(aq)$	
E		**F**	
$Cu(s) \rightarrow Cu^{2+}(aq) + 2e^-$		$Cu^{2+}(aq) + 2e^- \rightarrow Cu(s)$	

a) Identify the equation which shows iron(II) ions being oxidised.

b) Identify the two equations which show the reactions which occur when an iron nail is placed in copper(II) sulphate solution.

BITESIZEchemistry

Making electricity

Metals and corrosion

(SQA Topics: 11 & 12)

📺 ◉ This section is about:

- properties and uses of metals and alloys

- extracting metals

- reactions of metals

- corrosion

- protecting iron.

Over three-quarters of the elements are metals. The properties of metals include strength, conduction of heat and electricity, malleability and density, all of which can be put to good use in our everyday lives. These properties may be improved by alloying the metals with other elements. Steel is an alloy of iron and carbon and is one of our most widely used alloys.

Metals are found in the Earth's crust. Very unreactive metals like gold are found in the pure form. Most metals, however, are found as compounds known as ores. The metal has to be extracted from its ore, usually the metal oxide. Oxides of unreactive metals can be broken up by heating (decomposition). Oxides of more reactive metals have to be mixed with carbon then heated. Oxides of very reactive metals have to be melted then electrolysed. Iron is one of the most important metals as it is used to make steel. The iron is extracted from its oxide in a blast furnace. Carbon monoxide is passed over the heated oxide producing iron and carbon dioxide.

The reactivity of a metal explains why some metals are easier to extract than others. All metals react with oxygen to form the metal oxide but metals like sodium and magnesium are much more reactive than lead and silver.

Reactive metals like potassium and calcium react vigorously with water producing hydrogen gas, whereas metals like copper and gold do not react. Magnesium and zinc react with acids producing hydrogen gas but metals like copper and silver do not react. Results of these experiments and others allow us to put metals in order of how reactive they are. The most reactive metals are the most difficult to extract from their ore whereas the least reactive are the easiest. The dates of discovery of metals matches their reactivity. Very unreactive metals have been known for thousands of years whereas some very reactive metals have only been extracted on a large scale since the end of the 19th century.

All metals corrode; this means they react with water and oxygen from the air to form compounds. The metal atoms lose electrons and form ions. In the case of iron, the process is given the special name of 'rusting'. Ferroxyl indicator can be used to detect rusting. Iron is our most widely used metal and has to be protected from corrosion. Iron objects can either be coated with materials like paint, grease and other metals to stop the air and water reducing them, or connected to metals which will supply electrons to the iron and so slow down the rusting.

FactZONE

9 When you have finished this section, you should know the following at General Level:

- Metals have properties like heat and electrical conduction, strength, malleability and density which we can put to good use. The desirable properties of metals can be improved by alloying. Brass, solder and 'stainless' steel are alloys.

- Only very unreactive metals are found uncombined in the Earth, e.g. gold.

- Reactive metals have to be extracted from ores. The method of extracting a metal depends on its reactivity. Iron is extracted from its ore in a blast furnace. The two key reactions which take place in the blast furnace are the production of carbon monoxide and the reaction of this carbon monoxide with iron oxide to produce iron.

- There are social and industrial factors which have resulted in the large-scale extraction of more reactive metals.

- Metals react differently with oxygen, water and acid, depending on their reactivity. The reactivity of metals closely matches their dates of discovery – the least reactive were discovered first.

- Corrosion is a chemical reaction which takes place at the surface of a metal. Metals corrode at different rates. The corrosion of iron is known as rusting. Both oxygen and water are needed for rusting to occur. The first stage in rusting is when iron atoms lose two electrons to form $Fe^{2+}(aq)$ ions.

- Ferroxyl indicator can be used to show the extent of rusting. Ferroxyl indicator changes from green to blue in the presence of

$Fe^{2+}(aq)$ ions. Rusting is speeded up by the presence of an electrolyte such as the salt spread on the roads in winter.

- There are electrical and physical ways to protect iron from rusting. Electrical methods involve electrons being given to the iron. Cathodic and sacrificial protection are the two electrical methods. Physical methods create a barrier between the iron and air/water and include painting, greasing, coating in plastic, electroplating, tin-plating and galvanising.

C In addition, you should know the following at Credit Level:

- Extracting metal from its ore is an example of reduction. Reactive metals are more difficult to extract because the metal ion holds on to the oxide tightly.

- Rusting takes place in two stages – each is an oxidation process:
  $Fe(s) \rightarrow Fe^{2+}(aq) + 2e^-$ then
  $Fe^{2+}(aq) \rightarrow Fe^{3+}(aq) + e^-$

- The electrons lost by iron are gained by water and oxygen to form hydroxide ions.

- The loss of electrons taking place when iron rusts can be shown by setting up a chemical cell with iron and carbon. The blue colour around the iron electrode in a cell indicates rusting and the pink colour indicates that the metal is being protected.

- Dissolved carbon dioxide or another electrolyte is required, in addition to oxygen and water, before rusting occurs.

- When tin-plating is scratched and iron is exposed, rusting is speeded up because iron gives tin electrons.

Properties of metals/alloys

 Properties and uses

Metals have a number of desirable properties which puts them amongst our most useful substances:

- strength
- conduction of heat
- conduction of electricity – in both solid and liquid state
- density – some are lightweight
- malleability – they can be shaped.

These properties are put to good uses: copper is used in pots and pans because copper is a good conductor of heat, and in electrical flexes because it is a good conductor of electricity.

Aluminium is used for aircraft bodies because of its low density (lightweight).

Steel is used to make the girders and ropes which hold up the Forth Road Bridge because it is so strong. It is also used for car bodywork because it is strong and **malleable** – it can be rolled thin and pressed into different shapes.

◎ *Complete the summary table below to show the link between properties and use.*

Metal	Use	Property
	pots and pans	
		electrical conductor
aluminium		
	girders and ropes	
		malleability

Alloys

REMEMBER
Alloys are metals that are mixed with other elements. The properties of alloys are more useful than the metal on its own.

Although pure metals have useful properties, some properties can be improved by mixing metals with other elements. This is called alloying.

Steel is an alloy of iron with small amounts of other elements added to improve its strength and resistance to rusting. On its own, iron is very brittle, lacks strength and breaks fairly easily, but when small amounts of carbon are added it becomes much stronger, flexible and malleable. Stainless steel is iron with transition metals added. It does not rust whereas ordinary steel rusts quite quickly. Cutlery is made from stainless steel.

Duralumin is an alloy of aluminium with small amounts of copper added. Although aluminium is useful because it is lightweight, it lacks strength. Duralumin's properties of strength combined with low density makes it ideal for use as the bodywork of an aircraft.

Solder is an alloy of lead and tin. It has a lower melting point than either element which makes it useful for connecting components in electrical circuits.

Brass is an alloy of zinc and copper. Brass can be used for door handles and letter boxes.

Practice questions

1 a) Copper and tin are melted together to form bronze. What type of substance is bronze?

   b) Copper is used to make the bottoms of pots and pans. Which property of copper makes it suitable for this use?

2 A light bulb is made from different substances.

A	glass
B	argon
C	tungsten
D	molybdenum
E	brass
F	lead

Identify the substance which is an alloy.

3 Steel for different uses can be made by adding different metals to iron. Titanium is added to make steel suitable for use in rockets. Magnets are made from a steel which contains cobalt. The steel for railway lines is made by adding manganese to the iron. Drill bits require a very hard steel. This is made by adding tungsten to the iron.

Present the above information in a table with suitable headings.

C 4 The uses of metals are related to their properties.

Metal	Density (g/cm³)	Relative strength	Relative electrical conductivity
Aluminium	2.7	1.0	3.8
Steel	7.9	4.0	1.0
Copper	8.9	2.5	5.9

Overhead electricity cables have a steel core surrounded by an aluminium sheath.

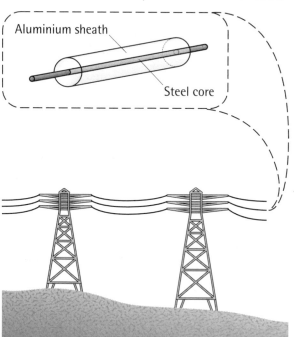

Aluminium sheath

Steel core

Using information from the table, suggest:

a) an advantage of using aluminium rather than copper for the cable.

b) why the cables have a steel core.

Extracting metals

📺 🔊 Metals are found in the ground but only very unreactive metals are found in their pure state. Gold is one example. Most metals are found as naturally occurring compounds known as ores, which are mainly metal oxides. Common ores are haemetite – an ore of iron, and bauxite – an ore of aluminium. Metals have to be extracted from their ores and there are several ways of doing this.

Heat alone

Metals like silver and mercury can be extracted by heating the metal oxide.

$$Ag_2O(s) \rightarrow Ag(s) + O_2(g)$$

C $$2Ag_2O(s) \rightarrow 4Ag(s) + O_2(g)$$

Heat and carbon

Iron is extracted from its ore in a blast furnace. Coke is made from coal and contains lots of carbon.

Raw materials 1

4 & 5

3
slag

6

hot air   hot air 2

molten iron

1	The raw materials (limestone, coke, iron ore) are put into the blast furnace.
2	The 'blast' is hot air, and the important part is oxygen.
3	Air reacts with coke carbon + oxygen → carbon dioxide
4	Carbon dioxide reacts with more coke. carbon dioxide + carbon → carbon monoxide
5	Iron ore + carbon monoxide → iron + carbon dioxide
6	Limestone reacts with impurities to produce 'slag'.

> ❗ **REMEMBER**
> Metals can be extracted from the ore by: heat alone, heat and carbon, or electrolysis.

The equations for the important reactions happening at 3, 4 and 5 are:

At (3): $$C(s) + O_2(g) \rightarrow CO_2(g)$$

At (4): $$CO_2(g) + C(s) \rightarrow CO(g)$$

C $$CO_2(g) + C(s) \rightarrow 2CO(g)$$

At (5): $$Fe_2O_3(s) + CO(g) \rightarrow Fe(s) + CO_2(g)$$

C $$Fe_2O_3(s) + 3CO(g) \rightarrow 2Fe(s) + 3CO_2(g)$$

The summary table on page 57 shows that there are some metals which cannot be extracted by chemical means. They have to be melted and a direct current (d.c.) passed through the melt. The positive metal ions are attracted to the negative electrode where they gain electrons and form metal atoms. Aluminium is extracted from bauxite in this way:

$$Al^{3+}(l) + 3e^- \rightarrow Al(s)$$

ion   atom

Summary

Metal oxide	Heat alone	Heat + carbon/carbon monoxide
sodium	No reaction	No reaction
magnesium		
aluminium		
zinc		Metal + carbon dioxide formed
iron		
tin		
lead		
copper		
mercury	Metal + oxygen formed	
silver		
gold		

C   In all three methods of extracting a metal from its ore, metal ions gain electrons and are changed into metal atoms.

Silver: $Ag^+ + e^- \rightarrow Ag$

Copper: $Cu^{2+} + 2e^- \rightarrow Cu$

Aluminium: $Al^{3+} + 3e^- \rightarrow Al$

Gain of electrons is known as **reduction**.

C   The ease of extraction of a metal from its ore is related to how reactive the metal is. The more reactive the metal, the more able it is to hold onto the oxygen, so the harder it is to break the attraction between the metal ion and the oxide ion.

Metal ores are finite resources – they will eventually run out. Recycling not only saves wasting metal and conserves our resources, it also saves energy.

Practice question

Many elements are metals.

A		B		C	
	zinc		copper		silver
D		E		F	
	tin		iron		potassium

a) Identify the metal produced in a blast furnace.

b) Identify the metal which is extracted from its ore by electrolysis.

c) Identify the metal which is extracted from its ore by heat alone.

Reactions of metals

(TV) (CD) The varying degree of difficulty in extracting a metal from its ore is due to the fact that different metals have different **reactivity**. This can be illustrated through a number of experiments.

Various metals when heated and put into oxygen react to form the metal oxide.

Metal and oxygen

Metal	Reaction with oxygen
Magnesium	very bright flame
Iron	glows brightly
Copper	dull glow

Metal and water

Metal	Reaction with water
Sodium	very reactive – bubbles of gas given off and metal bursts into flame
Magnesium	no reaction
Gold	no reaction

Metal and acid

Metal	Reaction with acid
Magnesium	lots of bubbles produced
Copper	no reaction
Silver	no reaction

! REMEMBER High reactivity indicates difficult extraction, whereas low reactivity indicates easy extraction.

Combining these results and the results of other experiments allows us to put the metals in order of how reactive they are, the most reactive at the top. This is called the **reactivity series**. It matches the order of the metals in the table in Extracting Metals on page 57.

This difference in reactivity of metals explains why the difficulty of extraction of metals varies.
- Very unreactive metals like gold are found uncombined.
- Unreactive metals like mercury can be extracted simply by heating the ore.
- Reactive metals like iron are obtained by mixing the ore with carbon and heating it.
- Very reactive metals like sodium are extracted by melting the ore and passing electricity through it.

Discovering metals

The differing ease of extraction of various metals matches their discovery dates. Gold which can be mined directly out of the ground was known in prehistoric times, whereas sodium, which requires electricity for extraction, was only extracted from its ore on a large scale relatively recently.

Discovery date	Metal	Reactivity
prehistoric times ↓ modern times	gold copper iron zinc sodium	unreactive ↓ very reactive

Social and industrial factors have resulted in the large-scale extraction of more reactive metals, e.g. the demand for iron (steel) during the Industrial Revolution in the nineteenth century and the industrial use of electricity in the twentieth century.

Practice questions

1   Many elements are metals.

A lead	B magnesium	C potassium
D silver	E tin	F zinc

   a) Identify the metal which can be found uncombined in the Earth's crust.

   b) Identify the most reactive metal.

   c) Identify the metal which does **not** react with acid.

2   Explain why sodium and potassium are stored under oil, but zinc and copper don't have to be.

G 3 The results of experiments carried out on four metals P, Q, R and S are shown below.

   i)   Metals Q and R burn in oxygen with intense flames. Metal S only glows when heated in oxygen.
   ii) Metals Q and S react with hydrochloric acid to give off hydrogen gas. Metal P does not react with hydrochloric acid.
   iii) Metal R reacts vigorously with water, but Metal Q reacts only slowly with water.

   a) Place the metals in order of reactivity, most reactive first.

   b) Suggest names for each of the four metals.

Corrosion

📺 ⊚ Metals have many and varied uses, but one problem is that they **corrode**. Corrosion is a chemical reaction at the surface of a metal, resulting in the formation of compounds. The rate of corrosion depends on the metal. Sodium corrodes very quickly when exposed to the air. Silver tarnishes very slowly; it is slowly corroding.

When iron (steel), our most widely used metal, corrodes it is given the special name **rusting**. Iron loses its strength when it rusts. Both oxygen (from the air) and water are needed before rusting occurs. This can be shown by the experiment set up below:

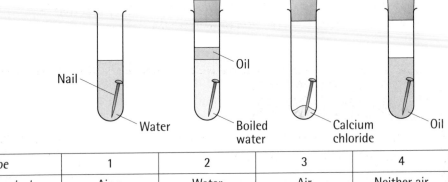

Tube	1	2	3	4
Chemical environment	Air + water	Water only	Air only	Neither air nor water
Amount of rust	Lots	None	None	None

Rust has a distinctive brown/orange colour and easily flakes off a piece of iron exposing more metal below.

In the first stage of rusting an iron atom loses two electrons and forms an $Fe^{2+}(aq)$ ion.

$$Fe(s) \rightarrow Fe^{2+}(aq) + 2e^-$$

Rusting can be indicated by the use of ferroxyl indicator. In the presence of $Fe^{2+}(aq)$ ions the indicator turns from green to dark blue. In the presence of $OH^-(aq)$ ions the indicator turns pink.

Rusting can be speeded up if an electrolyte is present. Salt is added to the roads in winter to stop them icing over. The salt dissolves in the water and ions are released. This then comes in contact with the underside of car bodywork. Any exposed metal will rust more quickly than it normally would.

© Rusting actually takes place in several stages, each involving oxidation (loss of electrons).

1st Stage: $Fe(s) \rightarrow Fe^{2+}(aq) + 2e^-$

2nd Stage: $Fe^{2+}(aq) \rightarrow Fe^{3+}(aq) + e^-$

The electrons lost by the metal during oxidation are gained by the water and oxygen to form hydroxide ions, $OH^-(aq)$ – this is **reduction**.

❗ REMEMBER The ion–electron equations for the oxidation and reduction processes happening during rusting are found on page 7 of the SQA Data Booklet. You don't have to memorise them!

$2H_2O(l) + O_2(g) + 4e^- \rightarrow 4OH^-(aq)$

C The oxidation and reduction processes taking place during rusting can be shown by setting up a chemical cell.

The meter shows the direction of electron flow through the leads from the iron to the carbon. The blue colour around the iron electrode shows that $Fe^+(aq)$ ions are being produced.

$Fe(s) \rightarrow Fe^{2+}(aq) + 2e^-$

The pink colour around the carbon electrode shows that $OH^-(aq)$ ions are being produced. Water and dissolved oxygen gain electrons from the electrode – the electrons which have come from the iron.

$2H_2O(l) + O_2(g) + 4e^- \rightarrow 4OH^-(aq)$

In addition to oxygen and water, dissolved carbon dioxide and other electrolytes are needed before rusting will occur.

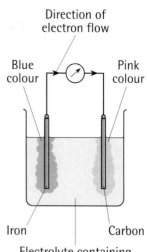

Direction of electron flow

Blue colour | Pink colour

Iron | Carbon

Electrolyte containing ferroxyl indicator

Practice questions

1 Steel ships can suffer from corrosion. Painting helps to prevent this.

   a) Why does painting help to prevent corrosion?

   b) Describe another way in which the ship could be protected from corrosion.

   c) Why is corrosion greater in sea water than in fresh water?

2 Aileen was investigating the rusting of iron. She set up four test tubes each containing a clean iron nail.

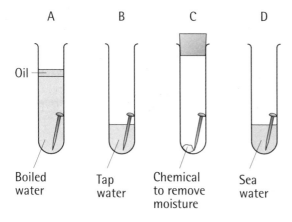

A   B   C   D

Oil

Boiled water | Tap water | Chemical to remove moisture | Sea water

Tube	Observations after 1 week
A	Nail still bright
B	Nail rusted
C	Nail still bright
D	Nail badly rusted

   a) Suggest why the nail in tube A did not rust.

   b) Why did the nail in sea water rust more than the nail in tap water?

C 3 When iron corrodes, iron atoms form Fe^{2+} ions. Fe^{2+} ions can oxidise further.

   a) Write the formula for the iron ion formed when Fe^{2+} ions oxidise further.

   b) Name the indicator which can be used to identify Fe^{2+} ions.

   c) State the colour change which occurs with the indicator in (b).

Protecting iron

There are two methods of protecting iron against rusting:

(1) Give electrons to the iron – **electrical** methods.

(2) Stop oxygen and water reaching the iron – **physical** (barrier) methods.

Electrical methods

When iron rusts it loses electrons, so if electrons are given to the iron, rusting should be slowed down.

a) Cathodic protection

The **cathode** is a name given to the negative side of a battery. The object to be protected is connected to the negative side of a battery and electrons flow to the object. On a large scale this is used on oil rigs and ships in dock.

b) Sacrificial protection

A metal higher in the electrochemical series is connected to one lower. Electrons flow from the higher metal to the lower one. Magnesium is higher than iron so electrons will flow to the iron when the two metals are connected. This is used to protect underground pipes.

If the iron was connected to a metal lower in the electrochemical series, the electrons would flow from the iron and rusting would be speeded up.

C The direction of electron flow in sacrificial protection can be shown in cells:

The pink colour around the iron shows that it is being protected. The electrons flowing through the wire from the magnesium are gained by the water and dissolved oxygen. Hydroxide ions OH$^-$(aq) are formed which turn the indicator pink.

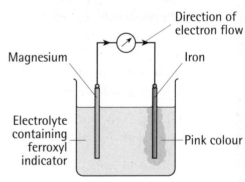

The blue colour around the iron shows that Fe^{2+}(aq) ions are being produced, i.e. the iron is rusting. The pink around the copper shows that OH$^-$(aq) ions are being produced. The water and dissolved oxygen are gaining the electrons which have come from the iron.

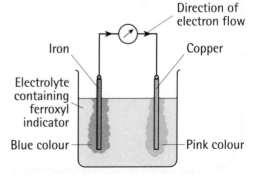

Physical methods

This involves providing a surface barrier between the metal and the air and water.

Painting – car bodywork, railings.

Greasing – moving parts on a bike, farm machinery.

Plastic coating – drying racks for wet crockery.

Electroplating – a layer of metal, such as silver, is deposited on another metal by electrolysis. The object to be plated is made the negative **electrode** in a solution containing ions of the metal being **deposited**. For silver plating: $Ag^+(aq) + e^- \rightarrow Ag(s)$.

Galvanising – iron objects are dipped into molten zinc. The zinc coating is quite tough, but even if it is scratched, the iron underneath is protected because electrons flow from zinc to iron because zinc is higher than iron in the electrochemical series. Car bodywork and nails can be protected by galvanising.

Tin plating – the insides of cans containing food can be plated with tin to stop corrosive juices attacking the can which is made from thin steel (iron).

C Tin is a soft metal which is easily scratched. Unlike zinc, when the iron is exposed the tin does not protect it because tin is lower in the electrochemical series than iron. Electrons flow from the iron to the tin so rusting is speeded up.

> **! REMEMBER**
> Iron reacts with water and oxygen to produce rust. This reaction can be prevented by coating the iron so that the reactants can't come into contact with one another.

Practice questions

1   Underground petrol storage tanks can be made of steel. The tanks can be protected from corrosion by attaching scrap metal to them.

(a) Which two substances must be present for steel to corrode?

(b) Explain fully why scrap magnesium can be used to protect the tanks but scrap copper cannot be used.

C 2   It is important to protect iron from rusting.

Identify the correct statement(s) about the rusting of iron.

A	Copper gives sacrificial protection to iron.
B	Ferroxyl indicator turns pink in the presence of Fe^{2+} ions.
C	Electroplating provides a surface barrier to air and water.
D	Tin plated iron rusts quickly when the coating is scratched.
E	Iron rusts when attached to the negative terminal of a battery.

Plastics

(SQA Topic: 13)

📺 🔊 This section is about:

- making plastics

- using plastics.

All plastics are made up of giant molecules called polymers. Polymers are made from thousands of small unsaturated molecules, generally called monomers, joined together. The process is known as addition polymerisation. The small unsaturated molecules are obtained by cracking long-chain alkanes found in the residue fraction of crude oil. Plastics have both common and chemical names. Polythene has the chemical name poly(ethene) – the part in brackets is the name of the alkene. Poly(ethene) is formed when the double bond in thousands of ethene molecules breaks allowing the monomers to add together. Many of our most common plastics are made by addition polymerisation, e.g. polystyrene and PVC.

Plastics have a number of useful properties – strength, heat and water resistance, electrical insulation, they are easily shaped, hardwearing and lightweight. These properties are reflected in their uses which range from polythene bags to bullet-proof vests! Some polymers are described as thermoplastic – they can be melted and reshaped. They have the advantage that they can be made in one factory and transported to another to be melted and shaped. Thermosetting polymers do not melt when heated. When the polymer is made, it has to be shaped straight away. Electrical plugs and sockets are made from thermosetting polymers because they won't melt if the plug overheats.

Most plastics are non-biodegradable – they won't rot. This is a big advantage when they are used as outdoor fittings on a house say, or as bumpers on a car. This property does, however, cause environmental problems. When plastic objects are thrown away they don't rot and thus pollute our environment – count the empty crisp bags and plastic bottles in your school playground! Most plastics burn or smoulder giving off toxic fumes. Many people are killed by inhaling the fumes created by burning plastics in house fires.

Natural polymers have useful properties, e.g. wool is warm and cotton is soft to touch. Synthetic fibres are often stronger than natural fibres. Poly(propene) for example can be woven into a material which can be used to lift objects which weigh several tons. Often synthetic and natural fibres are mixed to give a combination of strength and softness, e.g. polyester/cotton shirts.

ⓖ When you have finished this section, you should know the following at General Level:

- Plastics and synthetic fibres are examples of polymers. Polymers are giant molecules made from thousands of smaller molecules joined together.

- Synthetic means 'man-made'.

- The small unsaturated molecules used to make polymers are called monomers. Polymerisation is the name given to the process by which polymers are formed from monomers.

- Monomers are obtained by cracking large alkane molecules which come from the residue fraction of distilled crude oil. Polymers take their name from the monomer, e.g.

  chloroethene → poly(chloroethene)
     monomer   polymer

- How to use structural formulae to show polymerisation of ethene to form poly(ethene).

- Plastics have many useful properties:

  - durability (hardwearing), e.g. PVC water pipes
  - lightness, e.g. polythene bags
  - heat insulator, e.g. polystyrene cups
  - electrical insulator, e.g. electrical flex

- Thermoplastic polymers can be melted and reshaped.

- Thermosetting polymers cannot be melted and are suitable for making plugs, sockets and worktops.

- Non-biodegradable plastics do not rot and as a result can cause environmental pollution.

- Many plastics give off toxic fumes when burned.

- We make synthetic fibres from chemicals.

- Natural and synthetic fibres are both polymers. There are advantages and disadvantages to using natural and synthetic fibres, e.g. wool (natural) is soft and warm; kevlar (synthetic) is very flame-resistant.

ⓒ In addition, you should know the following at Credit Level:

- During polymerisation the C=C double bonds in the monomers break.

- The joining of unsaturated monomers is known as addition polymerisation.

- You should know how to construct a polymer given the structure of the monomer, e.g.

$$
\begin{array}{cccccc}
Cl & H & Cl & H & Cl & H \\
| & | & | & | & | & | \\
C & = C & C & = C & C & = C \\
| & | & | & | & | & | \\
H & H & H & H & H & H \\
\end{array}
$$

becomes

$$
\begin{array}{ccccccc}
 & Cl & H & Cl & H & Cl & H \\
 & | & | & | & | & | & | \\
- & C & - C & - C & - C & - C & - C & - \\
 & | & | & | & | & | & | \\
 & H & H & H & H & H & H \\
\end{array}
$$

- You should know how to recognise the repeating unit and monomer, given the structure of the polymer.

- The toxic gases given off when plastics burn depend on the elements in the plastic and include carbon monoxide, hydrogen chloride and hydrogen cyanide.

Making plastics

Plastics are examples of **polymers**. Polymers are giant molecules made by joining thousands of small unsaturated molecules, generally called **monomers**. The process is known as **polymerisation**.

Cracking

The small unsaturated molecules (monomers) used to make polymers are obtained by **cracking** large alkane molecules. These large molecules are found in the residue fractions obtained by the fractional distillation of crude oil.

$$\text{large alkanes} \quad \rightarrow \quad \text{small alkanes}$$

(For more information about cracking, see page 36.)

Naming polymers

Many of the polymers (plastics) used in everyday life have well-known names e.g. polythene, polystyrene, PVC. These are their common names. Their chemical name is taken from the name of the monomer. Polythene for example is made from ethene:

$$\text{ethene} \quad \rightarrow \quad \text{poly(ethene)}$$

(monomer)   (polymer)

The table below gives the chemical and common names of some polymers.

Monomer	Chemical name of polymer	Common name of polymer
ethene	poly(ethene)	polythene
chloroethene	poly(chloroethene)	PVC
phenylethene	poly(phenylethene)	polystyrene

Note that the chemical name for the polymer is simply the monomer name in brackets with the prefix 'poly'.

Note also how each monomer has 'ethene' in its name showing that all monomers have an ethene-type structure, i.e. a C=C double bond.

Name the polymer formed from propene.
Name the monomer used to make poly(tetrafluoroethene)

REMEMBER A **monomer** is a small unsaturated molecule. A **polymer** is a giant molecule formed when thousands of monomers join together. **Polymerisation** is the name given to the process by which monomers form polymers.

How poly(ethene) is formed

The formation of poly(ethene) can be shown by drawing the structural formula of three ethene molecules.

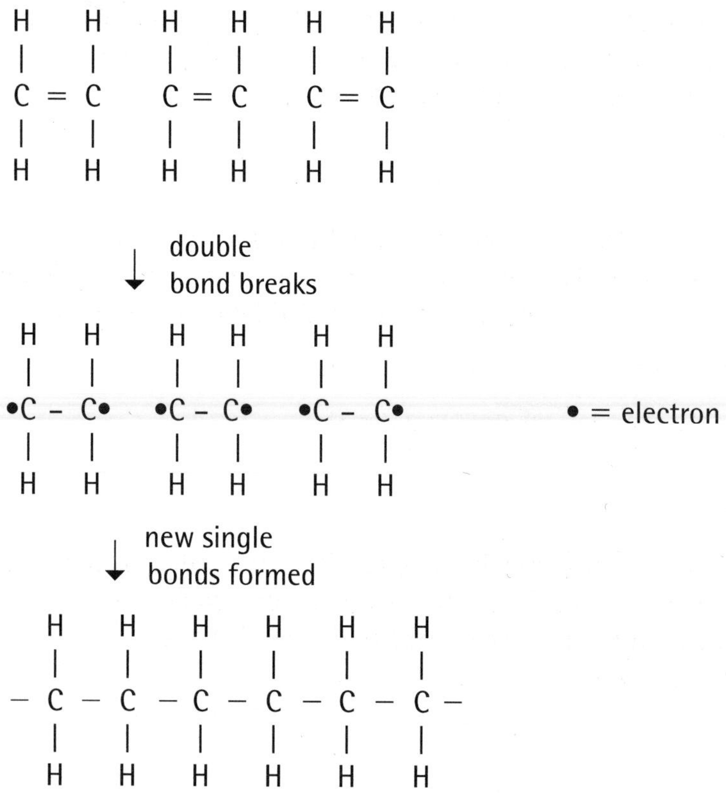

```
H   H   H   H   H   H
|   |   |   |   |   |
C = C   C = C   C = C
|   |   |   |   |   |
H   H   H   H   H   H
```

↓ double
   bond breaks

```
H   H   H   H   H   H
|   |   |   |   |   |
•C – C•  •C – C•  •C – C•        • = electron
|   |   |   |   |   |
H   H   H   H   H   H
```

↓ new single
   bonds formed

```
H   H   H   H   H   H
|   |   |   |   |   |
– C – C – C – C – C – C –
|   |   |   |   |   |
H   H   H   H   H   H
```

part of the poly(ethene) structure

G Formation of other polymers

Given the structural formula of any monomer, the structure of the polymer can be worked out. When drawing out the structure of the monomers draw them in a ⊟ shape to make sure the double bond is shown clearly as this is where the electrons which form the new bonds between the monomers came from.

Example:

When showing propene (C_3H_6) monomers joining to form poly(propene), draw the monomers like:

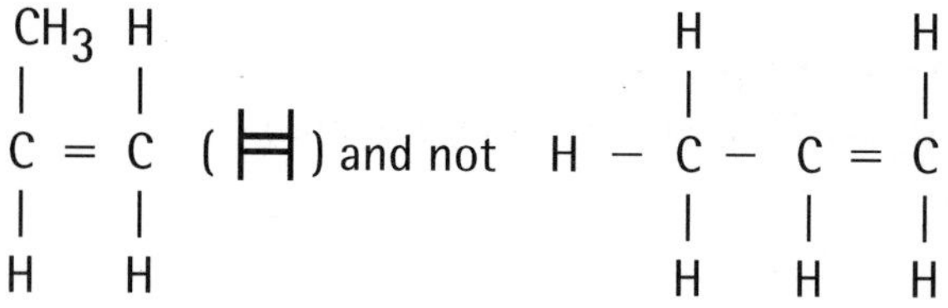

```
CH3 H                      H       H
|   |                      |       |
C = C  ( ⊟ ) and not   H – C – C = C
|   |                      |   |   |
H   H                      H   H   H
```

REMEMBER
The number of monomer units used to show polymerisation is usually three. Make sure you make it clear that each of the end carbon atoms is bonded to other monomers.

How poly(propene) is formed

```
CH3  H    CH3  H    CH3  H
 |   |     |   |     |   |
 C = C     C = C     C = C
 |   |     |   |     |   |
 H   H     H   H     H   H
                ↓
CH3  H         CH3  H         CH3  H
 |   |          |   |          |   |
•C = C•        •C = C•        •C = C•
 |   |          |   |          |   |
 H   H          H   H          H   H
                ↓
    CH3  H    CH3  H    CH3  H
     |   |     |   |     |   |
  –  C – C  –  C – C  –  C – C  –
     |   |     |   |     |   |
     H   H     H   H     H   H
```

part of the poly(propene) structure

REMEMBER Draw monomer units with the fololwing shape in order to show clearly how they join.

◎ *Show how three chloroethene monomer units join to form poly(chloroethene).*

```
 ⎧ Cl    H ⎫
 ⎪  |    | ⎪
 ⎨  C  = C ⎬
 ⎪  |    | ⎪
 ⎩  H    H ⎭
```

Given the structure of the polymer, the structure of the repeating unit and the monomer can be worked out. The structure of poly(propene) can be used as an example, with the part enclosed in the square bracket [ ] showing the repeating unit in the structure.

```
   CH3  H ⎡ CH3  H⎤ CH3  H
    |   | ⎢  |   |⎥  |   |
 – C – C ⎢– C – C⎥– C – C –
    |   | ⎢  |   |⎥  |   |
    H   H ⎣  H   H⎦  H   H

   CH3  H             CH3  H
    |   |              |   |
 – C – C  –         – C = C –
    |   |              |   |
    H   H              H   H

  Repeating unit        Monomer
```

Note the monomer structure can be worked out from the repeating unit – the monomer must have a C=C double bond.

◉ *From the structure of poly(phenylethene) shown below, work out the structure of the repeating unit and monomer.*

```
   C₆H₅  H     C₆H₅  H     C₆H₅  H
    |    |      |    |      |    |
 –  C  – C  –  C  – C  –  C  – C  –
    |    |      |    |      |    |
    H    H      H    H      H    H
```
$$\text{poly(phenylethene)}$$

Any chemical reaction in which the C=C double bond breaks is called **addition**. Polymerisation which involves the breaking of the C=C double bond is called **addition polymerisation**.

Practice questions

1   A bottle of sun lotion has the following label.

> **For recycling purposes:**
>
> bottle is polythene
>
> cap is polypropylene

G (a) Both polythene and polypropylene are made by addition polymerisation. What is meant by addition polymerisation?

G (b) The monomer for polyproplyene is propene.

```
   H              CH₃
    \            /
     \          /
      C   =   C
     /          \
    /            \
   H              H
```

Draw a part of the polypropylene chain which is formed by three monomer units linking together.

(c) How do we obtain monomers like propene?

2   During hip replacement operations the new hip join is fixed in place using bone cement.

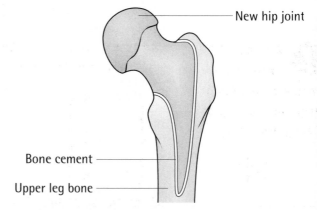

The most popular bone cement is a synthetic polymer formed from the monomer methyl methacrylate.

```
   H      CH₃
   |      |
   C  =   C
   |      |
   H      COOCH₃
```
methyl methacrylate

a) Name the polymer made from methyl methacrylate.

G b) Draw a section of this polymer, showing 3 monomer units joined together.

Using plastics

Advantages

Plastics have a number of properties which make them useful in our everyday lives:

- strength
- low density (lightweight)
- hardwearing
- heat resistant
- water resistant
- electrical insulators
- easily shaped.

The table below gives examples of plastics and uses which show some of their properties.

Plastic	Property	Use
polythene	lightweight and strong	'plastic' bags
polystyrene	heat insulator	cups for hot drinks
urea-formaldehyde	electrical insulator	plugs and sockets
melamine	hard wearing	kitchen worktops

Polythene and polystyrene are examples of plastics which can be melted and reshaped. They are **thermoplastic**.

Urea-formaldehyde and melamine are examples of plastics that cannot be melted. They are **thermosetting**. This is an advantage when they are used in electrical sockets and plugs as they won't melt if the plug overheats.

Most plastics are **non-biodegradable**. This means that they won't rot over the years. This is a particular advantage if the plastic is used outside where it is exposed to the weather.

Disadvantages

REMEMBER
Plastics are very useful because they are durable and don't rot, but this means that they present disposal and pollution problems.

Although being non-biodegradable has advantages, there are also disadvantages. Many so-called disposable items like food packaging and drinks bottles are plastic. When they are thrown away, they will not rot and, as a result, pollute the environment. Many of these plastics are thermoplastic and as such can be melted and reshaped. This means they could easily be recycled and made into other plastic objects like refuse sacks.

Many plastics burn or smoulder and give off toxic (poisonous) fumes in the process. A quick look around the kitchen at the number of plastic objects indicates the potential danger!

C The toxic gases given off during burning or smouldering depend on the elements present in the plastic.

Plastic	Elements	Toxic gas
polythene	C, H	carbon monoxide (CO)
PVC	C, H, Cl	hydrogen chloride (HCl)
polyurethane	C, H, N	hydrogen cyanide (HCN)

Natural versus synthetic

Plastics are **synthetic** polymers, i.e. we make them from chemicals. Natural polymers like wool and cotton originate from animals or plants.

There are advantages and disadvantages of both. Natural materials can be warmer and have a better look and feel next to the skin. However, they are not always hard-wearing. Natural materials like wood can rot whereas plastics do not.

Both synthetic and natural polymers can be made into fibres (threads). These fibres can be woven into materials which have a variety of uses. Synthetic fibres are often stronger than natural fibres and the two are often mixed together so that the resulting material is hardwearing as well as having a good feel next to the skin, e.g. polyester/cotton shirts. Some synthetic materials like nylon and terylene have the advantage that water evaporates quickly from their surfaces, and thus dry quickly when washed. Kevlar, one of our most modern synthetic plastics, is so strong that it is used as bullet-proof material and in the hulls of sea-going boats and canoes.

Practice questions

1   Alkathene is a polymer which is easy to melt. It is also resistant to the weather. Part of the structure of alkathene is shown.

```
   H   H   H   H   H   H
   |   |   |   |   |   |
 — C — C — C — C — C — C —
   |   |   |   |   |   |
   H   H   H   H   H   H
```

Identify the correct statement(s) about alkathene.

A It is biodegradable.
B It is a thermosetting polymer.
C It is produced when ethene is polymerised.
D It burns to produce carbon dioxide.

2
```
   H   H
   |   |
   C = C
   |   |
   H   CN (Acrylonitrile)
```

a) i)   Name the polymer made from acrylonitrile.

  ii)   The polymer can be melted and forced through tiny holes to form acrylic fibres. What term is used to describe a polymer which can be melted and reshaped?

b) Why are the fumes from burning plastics dangerous?

C c) Name the fumes given off when acrylonitrile burns.

Fertilisers

(SQA Topic: 14)

📺 This section is about:

- the need for fertilisers

- ammonia

- nitric acid.

The world's population is increasing fast. As a result, the demand for food is increasing, too. Plants are the source of all our food. They need nutrients to make them grow properly. The three essential elements in nutrients are nitrogen (N), phosphorous (P) and potassium (K). Plants take nutrients from the soil and they have to be replaced. Fertilisers are used to replace nutrients. There is not enough natural fertiliser like animal manure and compost to meet demand so synthetic fertilisers are used.

A good fertiliser must contain the essential elements and be soluble in water so that they can dissolve and be absorbed through the roots of plants. Farmers often use mixtures called NPK fertilisers because different crops need different amounts of each element. This high solubility plus overuse of fertilisers by some farmers is causing pollution problems.

In some parts of the country, fertilisers are being washed out of the soil and getting into the drinking water supply and polluting lochs. Some plants can fix nitrogen from the air. This means they can take atmospheric nitrogen and change it into nitrogen compounds which can be absorbed by the plant. These plants are known as legumes and include peas and clover.

Ammonia (NH_3) is one of the most important chemicals used in the manufacture of synthetic fertilisers. It is made industrially by a process called the Haber Process. Nitrogen and hydrogen are passed over a heated iron catalyst at high pressure. Ammonia is a colourless gas which is very soluble and forms an alkaline solution.

Nitric acid (HNO_3) is also important in the manufacture of fertilisers. It is made by dissolving nitrogen dioxide and air in water. Nitrogen dioxide is difficult to make because of the strong bonding between the atoms in a nitrogen molecule. It is formed in the air during lightning storms and inside a petrol engine. Nitrogen dioxide in the atmosphere contributes to acid rain. Acid rain can be a source of nitrogen for the soil but causes pollution problems. In each case, a high energy spark is passed through the air. This is not an economical way of making nitrogen dioxide on a large scale. Industrially, nitrogen dioxide is made by a process called the Ostwald Process. A mixture of ammonia and air is passed over a heated platinum catalyst.

g When you have finished this section, you should know the following at General Level:

■ The increasing world population has led to a need for more efficient food production.

■ Fertilisers are used to supply plants with nutrients containing the essential elements nitrogen, phosphorous and potassium. Natural fertilisers, like animal manure and compost, can supply these nutrients to plants.

■ The nitrogen cycle shows how nitrogen is supplied to, and taken out of, the soil. Nitrogen-fixing bacteria in the nodules of leguminous plants, like peas and clover, can convert atmospheric nitrogen into nitrogen compounds which plants can absorb.

■ Synthetic fertilisers must contain the essential elements and be soluble in water. Useful synthetic fertilisers include ammonium and potassium salts, nitrates and phosphates. Ammonium nitrate is a widely used fertiliser made by reacting ammonia with nitric acid.

■ Overuse of fertilisers can create pollution problems such as polluted drinking water and lochs.

■ Industrially, ammonia is made by reacting nitrogen with hydrogen and is known as the Haber Process. The requirements for the Haber Process are an iron catalyst, moderately high temperature and high pressure. Ammonia is a colourless gas with an unpleasant smell and is very soluble in water forming an alkaline solution.

■ Nitric acid is made by dissolving nitrogen dioxide in water. Nitrogen dioxide is formed in the air during lightning storms and in petrol engines.

■ Nitrogen gas is very unreactive because of the strong triple bond between the atoms in its molecule. When a spark is passed through air, it supplies sufficient energy to break the strong triple bond in a nitrogen molecule. However, sparking air is an uneconomical way to produce nitrogen dioxide. Nitrogen dioxide forms acid rain in the atmosphere. Acid rain supplies nitrogen to the soil but causes pollution problems.

■ Industrially, nitrogen dioxide is made by reacting ammonia with oxygen from the air – this is known as the Ostwald Process. In the Ostwald Process, ammonia and air are passed over a platinum catalyst at a moderately high temperature. The Ostwald Process can be demonstrated in the laboratory.

C In addition, you should know the following at Credit Level:

■ Bacterial methods of fixing nitrogen are cheaper than chemical methods.

■ Different crops need fertilisers containing different proportions of nitrogen, phosphorous and potassium.

■ The Haber Process is carried out at a moderately high temperature to stop the ammonia formed from breaking up. The formation of ammonia by reacting nitrogen and hydrogen is a reversible reaction. Ammonia can be made in the laboratory by heating an ammonium compound with an alkali.

■ In the Ostwald Process, the reaction is carried out at a moderately high temperature to stop the nitrogen oxide breaking up. The process is an exothermic reaction so the catalyst does not have to be continually heated.

Fertilisers

The need for fertilisers

The population of the world is constantly increasing. As the number of people increases so does our demand for more food and the need for efficient food production. To help increase the amount of food they can produce, farmers use fertilisers to provide plants with the nutrients they need for healthy growth. The main nutrients are **nitrogen, phosphorous** and **potassium**.

REMEMBER Artificial fertilisers are necessary to produce enough food for an expanding population.

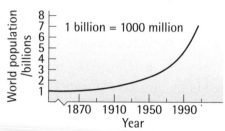

1 billion = 1000 million

World population /billions

1870   1910   1950   1990
Year

Last century, when the population was smaller, we could use natural fertilisers like horse manure and compost to provide the soil with the essential elements. Today, this just isn't practical for the amount of food we need to produce. Only organic farmers use natural fertilisers.

The nitrogen cycle

The **nitrogen cycle** diagram shows how nitrogen is naturally recycled but also how we are affecting this cycle. Plants and animals, before farming, would die then decompose and return nitrogen compounds to the soil to be used by plants in the following year. Intensive farming now means we take a lot of plants and animals out of the cycle so the nitrogen has to be replaced in other ways.

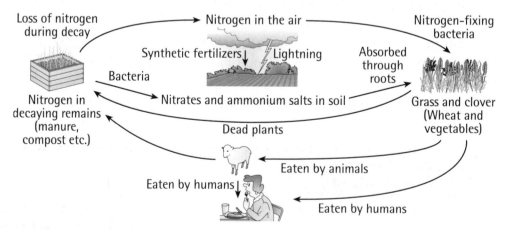

Loss of nitrogen during decay

Nitrogen in the air

Nitrogen-fixing bacteria

Synthetic fertilizers   Lightning

Absorbed through roots

Bacteria

Nitrogen in decaying remains (manure, compost etc.)

Nitrates and ammonium salts in soil

Grass and clover (Wheat and vegetables)

Dead plants

Eaten by animals

Eaten by humans

Eaten by humans

Leguminous plants

Plants which are **leguminous** have nodules on their roots which contain **nitrogen-fixing bacteria**. This means they can take nitrogen from the air and change into compounds that plants can use. This means they don't need fertilisers. Examples include clover, beans, peas and soya.

This bacterial method of fertilisation is cheaper than chemical methods.

Synthetic fertilisers

Most plants have to get their nutrients from the soil so farmers need to use synthetic fertilisers. Synthetic means we make them in factories. A good synthetic fertiliser must have two properties:
- it must contain the essential elements
- it must be soluble in water so that the plants can absorb the essential elements through their roots, in the form of a solution.

The following types of compounds are amongst our most common fertilisers: ammonium and potassium salts; phosphates and nitrates, e.g. potassium phosphate and ammonium nitrate.

Ammonium nitrate is one of the most widely used synthetic fertilisers. It is made by reacting ammonia with nitric acid:

$$\text{ammonia} + \text{nitric acid} \rightarrow \text{ammonium nitrate}$$

$$NH_3(g) + HNO_3(aq) \rightarrow NH_4NO_3(aq)$$

In industry, the water is evaporated from the nitrate solution and the solid fertiliser is sold in bags under the trade name NITRAM. It is spread on the fields by machine.

© Different crops need fertilisers containing different proportions of nitrogen (N), phosphorous (P) and potassium (K). Farmers use ready-mixed fertilisers called NPK fertilisers and select the composition to suit the crop they are growing.

Pollution problems

There is some concern that, in certain areas of the country, farmers are using too much synthetic fertiliser. Nitrate fertilisers are very soluble and, as a result, are easily washed out of the soil. They can **leak** into rivers and lochs and can end up in our water supply. Some doctors are concerned that too much nitrate can damage our health. They can also cause algae to form on the surface of the loch which reduces levels of oxygen in the water, so fish and other aquatic creatures die.

◎ *Make a list of the benefits and disadvantages of using artificial fertilisers.*

Practice questions

1 a) Name the essential elements plants need to ensure healthy growth.

   b) Explain how plantsi like beans and clover are able to 'fix' nitrogen from the air.

   c) How do plants which cannot 'fix' nitrogen obtain an adequate supply?

   d) Explain why there is increasing demand for more food to be produced.

2 a) Give two reasons why ammonium nitrate (NH_4NO_3) is a good fertiliser.

   b) Describe two ways in which fertilisers could pollute our environment.

Fertilisers

Ammonia

📺 Industrial production

Ammonia is made by reacting nitrogen and hydrogen – this is known as the **Haber Process.**

Nitrogen (from the air) and hydrogen (from the oil industry) are passed over an iron catalyst at a temperature of around 500°C and a pressure of 200 atmospheres.

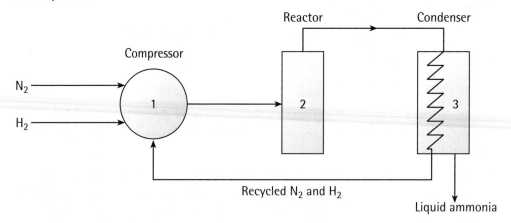

REMEMBER Ammonia contains nitrogen which is very important for plant growth.

C In the compressor (1) the pressure is at around 200 atmospheres. This forces the reactant molecules together so there is more chance of reaction. In the reactor (2), the gas mixture is passed over an iron catalyst at 500°C. The high temperature speeds up the chemical reaction but this reaction is reversible. This means that not only do nitrogen and hydrogen react, but the ammonia starts to break up to form reactants. Too high a temperature will cause the ammonia to break up so a moderate temperature is used. In the condenser (3), the ammonia gas is liquefied and drained off.

Any gas that does not react is recycled through the compressor and reactor.

$$\text{nitrogen} + \text{hydrogen} \rightleftharpoons \text{ammonia}$$
$$N_2(g) + 3H_2(g) \rightleftharpoons 2NH_3(g)$$

The ⇌ indicates the reaction gas in both directions, i.e. it is a reversible reaction.

REMEMBER The properties of ammonia are that it is a colourless gas, it has an unpleasant smell, it's very soluble and turns moist pH paper blue (alkaline).

Laboratory production

C Ammonia can be prepared in the laboratory by heating an ammonium compound with an alkali.

ammonium chloride	+	sodium hydroxide	→	sodium chloride	+	water	+	ammonia
$NH_4Cl(s)$	+	$NaOH(s)$	→	$NaCl(s)$	+	$H_2O(g)$	+	$NH_3(g)$

Nitric acid

📺 Nitric acid is made by dissolving a mixture of nitrogen dioxide and air in water.

nitrogen dioxide (brown gas) + air + water → nitric acid

Nitrogen dioxide is formed in the air during lightning storms – air is mainly nitrogen and oxygen. Nitrogen dioxide is also formed in a car engine when petrol burns in air. In both cases, a high energy spark causes nitrogen and oxygen in the air to combine.

$$\text{nitrogen} + \text{oxygen} \rightarrow \text{nitrogen oxide}$$

$$N_2(g) + O_2(g) \rightarrow NO_2(g)$$

© $\quad N_2(g) + 2O_2(g) \rightarrow 2NO_2(g)$

The nitrogen atoms are held together by a very strong triple bond and it requires a lot of energy to break this bond. This shows that nitrogen gas is not very reactive.

The nitrogen dioxide produced during lightning storms and in car engines dissolves in moisture in the atmosphere forming acid rain. The nitrogen in the acid rain helps replace nitrogen in the soil but also causes environmental problems.

Industrial manufacture

Sparking air is not an economical route to nitric acid. In the industrial manufacturing process, ammonia gas is mixed with air and passed over a hot platinum catalyst. Nitrogen oxide (NO) is formed which reacts with oxygen in the air to form nitrogen dioxide (NO_2). The nitrogen dioxide is mixed with more air and passed up an absorption tower which has water flowing down it. The gases dissolve forming nitric acid (HNO_3).

© The reaction of ammonia with oxygen is carried out at a moderately high temperature to stop the nitrogen oxide breaking up and reforming ammonia and oxygen.

© The reaction of ammonia with oxygen is **exothermic**. The heat from the reaction heats the catalyst, so once the reaction has started the external heat supply to the catalyst can be switched off.

The catalytic conversion of ammonia to nitrogen dioxide can be carried out in the laboratory.

The platinum catalyst is seen to glow red hot and a very faint brown colour is seen, showing the presence of nitrogen dioxide.

Fertilisers

Carbohydrates and related compounds

(SQA Topic: 15)

📺 🔊 This section is about:

- photosynthesis and respiration

- carbohydrates

- making and breaking carbohydrates

- alcoholic drinks.

Animals need carbohydrates as a source of energy. Carbohydrates are produced by green plants during photosynthesis. During photosynthesis, carbon dioxide and water combine in the presence of light and chlorophyll to produce glucose and oxygen. Glucose is broken down by animals and plants to produce energy. This is called respiration. Carbon dioxide and water are the chemical products. Photosynthesis and respiration maintain the balance of oxygen and carbon dioxide in the air. The destruction of rainforests and the production of carbon dioxide from the combustion of fossil fuels may upset this balance.

Carbohydrates are made up of carbon, hydrogen and oxygen. There are twice as many hydrogens as oxygens in each carbohydrate molecule. Some carbohydrates, like glucose, are sweet-tasting and dissolve in water. Others, like starch, are not sweet and don't dissolve in water. Carbohydrates look similar but can be told apart by chemical tests. Starch turns brown iodine a blue/black colour. Glucose turns pale blue Benedict's solution an orange/brown colour.

Carbohydrates can be classified as monosaccharides ($C_6H_{12}O_6$), disaccharides ($C_{12}H_{22}O_{11}$) and polysaccharides ($C_6H_{10}O_5$)n, where n = large number.

Thousands of glucose molecules join together in plants to form starch. This process is called condensation polymerisation. Starch itself is broken down by enzymes in our bodies and glucose molecules are formed. This process is called hydrolysis. Starch can be hydrolysed in the laboratory using enzymes or acid.

Alcoholic drinks contain ethanol, a member of the alkanol family. Ethanol is made by adding yeast to any fruit or vegetable containing starch or glucose. Yeast is a living organism and its enzymes can break down glucose into ethanol and carbon dioxide. The process is known as fermentation. The efficiency of an enzyme depends on temperature and pH. During fermentation, the ethanol reaches a concentration at which the enzymes are killed. In order to raise the concentration of ethanol, the fermentation mixture is distilled. The ethanol evaporates first, having the lower boiling point, and is condensed and collected.

⑨ When you have finished this section, you should know the following at General Level:

■ Carbohydrates supply the energy used to keep us warm and moving around. When carbohydrates burn, they release energy and carbon dioxide and water are produced.

■ Green plants make carbohydrate and oxygen during photosynthesis. During photosynthesis, carbon dioxide combines with water in the presence of light and chlorophyll. Chlorophyll absorbs light during photosynthesis.

■ Respiration is the process by which animals and plants obtain a supply of energy by breaking down glucose to give water and carbon dioxide. Photosynthesis in plants and respiration in animals is important in maintaining the balance of carbon dioxide and oxygen in the air.

■ Cutting down the rain forests could upset the balance of gases in the air.

■ Carbohydrates contain carbon, hydrogen and oxygen. Glucose, sucrose and starch are all carbohydrates. Glucose is sweet and dissolves in water. Starch is not sweet and does not dissolve in water. A beam of light shows up in a starch solution but not in a glucose solution.

■ Only starch turns brown iodine solution a blue/black colour. Glucose turns pale blue Benedict's solution an orange/brown colour when they are heated together.

■ Glucose is made during photosynthesis and polymerises to form starch. Digestion breaks down large starch molecules into small glucose molecules which can be absorbed into the blood. The enzyme amylase in saliva acts as a biological catalyst in the break down of starch.

The break down of starch can be carried out in the laboratory using enzymes or acid.

■ The alcohol in alcoholic drinks is called ethanol and is a member of the alkanol family. Alcoholic drinks can be made from the carbohydrate in any fruit or vegetable. It is the source of the carbohydrate that makes them different, e.g. apples–cider and grapes–wine. Alcoholic drinks are made by the fermentation of glucose. Yeast provides the biological catalyst for fermentation.

■ The concentration of alcohol in drinks can be made higher by distillation. Ethanol boils at 79°C and water at 100°C so they can be separated by distillation.

ⓒ In addition, you should know the following at Credit Level:

■ Glucose/fructose and maltose/sucrose are pairs of isomers.

■ The Benedict's test works for fructose and maltose as well as glucose.

■ When glucose polymerises to form starch, water molecules are produced – this type of polymerisation is known as condensation.

■ Hydrolysis is the breaking down of a molecule by the addition of water. Starch can be hydrolysed to glucose.

■ Monosaccharides have the formula $C_6H_{12}O_6$, e.g. glucose. Disaccharides have the formula $C_{12}H_{22}O_{11}$, e.g. sucrose.

■ Enzyme activity can be reduced by high temperature and changes in pH. During fermentation, the concentration of alcohol is limited because it kills the enzymes in yeast.

Carbohydrates and related compounds

Photosynthesis and respiration

Our body needs energy to keep us warm and enable us to move about. This energy comes from the food we eat that contains **carbohydrates**. Most sweet-tasting food contains carbohydrates called glucose and sucrose – these are commonly known as sugars. Starch is another carbohydrate found in potatoes, rice and pasta.

Photosynthesis

Green plants make carbohydrates. They take in **carbon dioxide** (CO_2) and **water** (H_2O), which are small molecules, and change them into larger, more complicated carbohydrate molecules. The process needs light. Green leaves in plants contain a chemical called **chlorophyll** which traps the light energy needed for the reaction. **Oxygen** gas is produced and released into the atmosphere. The whole process is called **photosynthesis**. **Glucose** ($C_6H_{12}O_6$) is the carbohydrate formed during photosynthesis.

Light   Oxygen

Chlorophyll   Glucose

Water

Carbon dioxide

$$\text{Carbon dioxide} + \text{water} \xrightarrow[\text{Chlorophyll}]{\text{Light}} \text{Glucose} + \text{oxygen}$$

$$CO_2(g) + H_2O(l) \longrightarrow C_6H_{12}O_6(aq) + O_2(g)$$
C $6CO_2(g) + 6H_2O(l) \longrightarrow C_6H_{12}O_6(aq) + 6O_2(g)$

Respiration

The energy we get from eating carbohydrates is released only after a series of reactions involving oxygen. These reaction take place in our body cells. Oxygen reacts with carbohydrate producing energy and the chemical products carbon dioxide and water. The whole process is known as **respiration**. Respiration also takes place in plants. The energy is needed for important reactions which help the plant cells to develop and multiply.

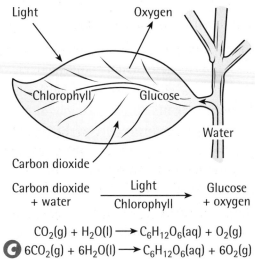

Carbon dioxide + water

Oxygen

Carbohydrate

Heat energy

Movement energy

If a carbohydrate is burned in the laboratory the chemical products, carbon dioxide and water, can be detected.

$$\text{Glucose} + \text{oxygen} \longrightarrow \text{carbon dioxide} + \text{water}$$

$$C_6H_{12}O_6(aq) + O_2(g) \longrightarrow CO_2(g) + H_2O(g)$$
C $C_6H_{12}O_6(aq) + 6O_2(g) \longrightarrow 6CO_2(g) + 6H_2O(g)$

C   The fact that carbon dioxide and water are produced when a carbohydrate burns shows that carbohydrates must contain carbon and hydrogen – there is no other source of these elements. Carbohydrate also contains oxygen, so they are **not** hydrocarbons.

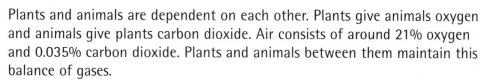

U-tube in ice water

Water collected

Burning sugar

Carbon dioxide turns limewater cloudy

To suction pump

Carbon dioxide in the atmosphere

A closer look at photosynthesis and respiration shows that the two are chemically opposite reactions:

Respiration: glucose + oxygen → carbon dioxide + water

Photosynthesis: carbon dioxide + water → glucose + oxygen

Plants and animals are dependent on each other. Plants give animals oxygen and animals give plants carbon dioxide. Air consists of around 21% oxygen and 0.035% carbon dioxide. Plants and animals between them maintain this balance of gases.

Human activity is thought to be upsetting this balance. The burning of fossil fuels is increasing the amount of carbon dioxide in the air, while the destruction of rainforests is reducing the amount of oxygen being supplied by trees and reducing the amount of carbon dioxide absorbed by them.

Some scientists fear that the increase in carbon dioxide is contributing to increased global warming; this is known as the **greenhouse effect**.

◎   *Draw a diagram that shows the mutual dependence of plants and animals and list the processes occurring on each side.*

Practice questions

1   Plants make glucose by photosynthesis.

   a) Name the gases taken in and given out during photosynthesis.

   b) To which family of compounds does glucose belong?

   c) What does chlorophyll do during photosynthesis?

2   Explain why photosynthesis and respiration are important processes for maintaining the balance of gases in the atmosphere.

3   Many chemical reaction involve gases.

A	B	C
ammonia	oxygen	hydrogen
D	E	F
nitrogen	carbon dioxide	nitrogen dioxide

   a) Identify the gas which is used up during photosynthesis.

   b) Identify the gas which is produced by respiration.

Carbohydrates and related compounds

Carbohydrates

Formulae

All carbohydrates contain the elements carbon, hydrogen and oxygen.

Some common examples are shown in the table below:

Name	Formula
fructose	$C_6H_{12}O_6$
glucose	$C_6H_{12}O_6$
maltose	$C_{12}H_{22}O_{11}$
sucrose	$C_{12}H_{22}O_{11}$
starch	$(C_6H_{10}O_5)n$

n = large number

Each formula shows that there are twice as many hydrogen atoms per molecule than oxygen atoms. This is the same for all carbohydrates.

C Some carbohydrates in the table have the same molecule formula but different names. This is because they have different structures and are called **isomers**.

C Carbohydrates are classified according to the number of carbon atoms in the molecule. The simplest carbohydrates have the formula $C_6H_{12}O_6$. They are **monosaccharides**. When two monosaccharides join, a **disaccharide** is formed. They have the formula $C_{12}H_{22}O_{11}$. When thousands of monosaccharides join, a **polysaccharide** is formed. They have the general formula $(C_6H_{10}O_5)n$ where n is a very large number which varies.

REMEMBER Carbohydrates have isomers but you don't have to be able to draw them! Fructose and glucose are isomers Maltose and sucrose are isomers.

The table below gives some examples:

Monosaccharides $(C_6H_{12}O_6)$	Disaccharides $(C_{12}H_{22}O_{11})$	Polysaccharides $(C_6H_{10}O_5)n$
fructose	maltose	starch
glucose	sucrose	cellulose

Properties of carbohydrates

Many foods and drinks are sweet-tasting. Often this is because they contain the sweet-tasting carbohydrates like **glucose** ,commonly known as sugars. Sugars are also soluble in water. Starch is not sweet-tasting and does not dissolve well in water. Solubility can be used to distinguish between a glucose solution and a starch 'solution'. To the naked eye, both carbohydrates appear to form colourless solutions, but when a beam of light is passed through both, it shows up clearly in the starch but not in the glucose, which is a true solution. The starch molecules are too big to totally dissolve and the light reflects off them.

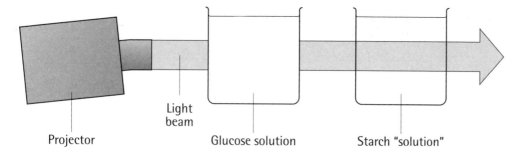

Projector   Light beam   Glucose solution   Starch "solution"

Chemical tests

Starch turns brown iodine solution a blue/black colour. As none of the other carbohydrates have this effect, this can be used as a test for starch.

Glucose turns Benedict's solution from pale blue to orange/brown in colour when they are warmed together. This can be used as a test to tell apart glucose and sucrose.

C The Benedict's test works for other carbohydrates like fructose and maltose so can't be used as a test to tell glucose apart from all other carbohydrates.

Practice questions

1 Bees collect nectar from flowers to make honey. The nectar contains sucrose which is broken down in the bees' bodies to give glucose and fructose.

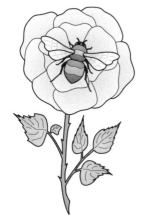

Composition of Clover Honey

Water	18%
Glucose	35%
Fructose	40%
Other compounds	7%

a) Name the family compounds to which glucose and fructose belong.

b) Name the type of substance in the bees' bodies which can break down sucrose.

c) Describe the test you would use to show that honey contains glucose.

2 Carbohydrates are very important foodstuffs.

A	fructose
B	glucose
C	maltose
D	starch
E	sucrose

a) Identify the **two** carbohydrates which do not give a positive result when tested using Benedict's or Fehling's Reagent.

b) Identify the **two** carbohydrates which have the molecular formula $C_6H_{12}O_6$.

c) Identify the carbohydrate which does not dissolve well in water.

Carbohydrates and related compounds

Making and breaking carbohydrates

Making starch

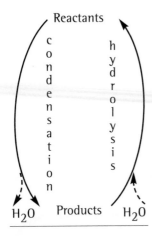

REMEMBER Condensation and hydrolysis are opposite reactions

Glucose is made in green plants during photosynthesis. Thousands of the glucose molecules join together to form starch. This is polymerisation. The glucose molecules are the **monomers** and starch is the **polymer**.

C Glucose has a complicated structures but we can simplify it to see what happens when glucose molecules join together.

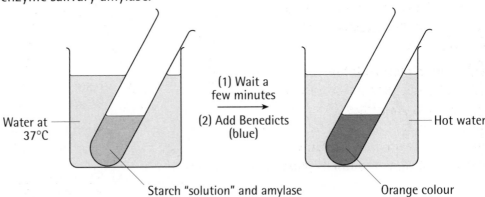

C Each time two glucose molecules join, a molecule of water is produced. Polymers formed in this way are known as **condensation polymers**.

Breaking starch

The food molecules we eat are broken down in our body. This is called **digestion**. Biological catalysts called **enzymes** digest the food. Starch is broken down by an enzyme in saliva called **salivary amylase**. The starch molecules are too big to get through the gut wall and into the blood stream so they are broken down into the smaller glucose molecules. Glucose molecules are small enough to be absorbed into the blood stream and then into our body cells where they react to produce energy.

The breaking down of starch can be done in the laboratory using acid or the enzyme salivary amylase.

C The breaking down of starch is called **hydrolysis**. When glucose polymerises to form starch, water molecules are produced. When starch is hydrolysed, glucose molecules are produced and water is added back on.

C Sucrose is hydrolysed in a similar way to form glucose and fructose.

Alcoholic drinks

Fermentation

The alcohol in alcoholic drinks is made from carbohydrate. Plants are our source of carbohydrate so almost any fruit or vegetable can be used to make alcoholic drinks. The type of plant used determines the final flavour of the drink. The table below gives examples of drinks and their plant source.

Drink	Source of carbohydrate	% alcohol
beer	barley	3–5
cider	apples	3–7
wine	grapes	9–13
whisky	barley	40

The alcohol in alcoholic drinks is called **ethanol** and is a member of the **alkanol** family. It is produced by a process called **fermentation**. During fermentation, glucose is broken down into ethanol and carbon dioxide. The process requires the presence of **yeast** which contains enzymes. The enzymes act as **biological catalysts**.

$$\text{glucose + water} \xrightarrow[\text{37°C}]{\text{yeast}} \text{ethanol + carbon dioxide}$$

G   Enzymes work very fast but their speed can be affected if the temperature is too high or the pH is not correct. The enzymes become denatured. This means their shape changes and they cannot act as a catalyst. Most enzymes in our body work best at body temperature which is 37°C.

G   During fermentation, the ethanol concentration increases as the reaction progresses. When the solution contains about 13% ethanol, the fermentation stops because the ethanol kills the enzymes in yeast.

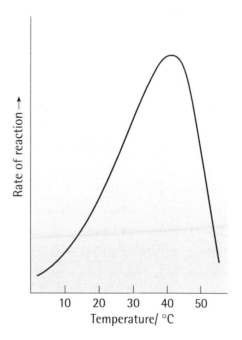

Distillation

A look at the table of alcoholic drinks shows whisky with 40% alcohol. This is only possible because the ethanol is separated out by distillation in order to increase its concentration. Water and ethanol can be separated because they have different boiling points – ethanol boils at 79°C and water at 100°C. When the mixture is heated, the ethanol evaporates first and is collected separately.

Carbohydrates and related compounds

Formulae, equations and calculations

(SQA Topic: permeates syllabus)

◉ This section is about:

- chemical formulae

- chemical equations

- chemical calculations.

Chemical formulae

The chemical formula of a compound, both covalent and ionic, indicates the ratio of atoms or ions present. In a covalent molecular compound, the chemical formula indicates the actual number of atoms in each molecule and is often called the molecular formula.

Formulae from valency

REMEMBER The name of a two-element compound ends in -ide.

Valency is another word for combining power. The valency of an element tells us how many atoms of another element a particular atom can combine with. The valency of a particular element can be worked out from its group number in the periodic table and is summarised in the following chart.

Group	Valency	Example
1–4	same as group number	K: group 1: valency = 1
		C: group 4: valency = 4
5–7	8 – group number	N: group 5: valency = $8 - 5 = 3$
		O: group 6: valency = $8 - 6 = 2$

The noble gases have a valency of O. Hydrogen has a valency of 1. The table below shows how valency can be used to work out the chemical formulae of simple two-element compounds.

REMEMBER The valencies are swapped in order to get the number of atoms of each element.

Elements	H	S	Mg	Cl	Al	O
Valency	1	2	2	1	3	2
No. of atoms	2	1	1	2	2	3
Formula	H_2S		$MgCl_2$		Al_2O_3	
Name	hydrogen sulphide		magnesium chloride		aluminium oxide	

BITESIZEchemistry

A formula shows the smallest ration of atoms, e.g. silicon dioxide.

Elements	Si	O
Valency	4	2
No. of atoms	2	4
Formula	Si_2O_4 =	SiO_2

Exceptions to the rule

The formulae of some compounds can't be worked out using the valency method, but their name often has a prefix to indicate the ratio of atoms.

mono- = 1,   di- = 2,   tri- = 3,   tetra- = 4 etc.

If there is no prefix then there is only one atom of that element.

Example 1: nitrogen dioxide – NO_2 2: phosphorous trichloride – PCl_3

Elements

Most elements have their symbol as their formula. The exception are the diatomic molecules, e.g. H_2

Ions containing more than one kind of atom

Some compounds have ions containing more than one kind of atom. These are found on page 4 of the SQA Data Booklet. Note that most of the negative ions have names ending in -ate or -ite. The number of charges on each ion tells you its valency, so the valency method can be used to work out the formulae.

Example	sodium	carbonate
	Na	CO_3^{2-}
Valency	1	2
Swap	2	1
Formula	Na_2CO_3	

The change on the ion is not usually shown in the formula.

G If you are asked to show the charges on ions in formulae, both ions must be bracketed. For sodium carbonate: $(Na+)_2 (CO_3^{2-})$

G Roman numerals

Some ionic compounds have Roman numerals in their names. This number tells you the valency of the ion in the first part of the name.

! REMEMBER
You need to know the formulae of the diatomic elements:

Hydrogen	H_2
Nitrogen	N_2
Oxygen	O_2
Fluorine	F_2
Chlorine	Cl_2
Bromine	Br_2
Iodine	I_2

Example   silver(I) oxide   Ag   O

            valency   1   2

            swap   2   1

            formula Ag_2O

⊙ Brackets

Brackets are used when there is more than one ion with more than one atom in a formula.

Example   copper(II) nitrate   Cu NO_3^-

            valency   2   1

            swap   1   2

            formula $Cu(NO_3)_2$

Chemical equations

REMEMBER Always use the valency method when working out the formulae of compounds to be used in the equation.

A chemical equation is a short-hand way of showing chemicals reacting and chemicals being produced.

reactants   →   products

Chemical equations can be written in words and formulae.

Example 1:   Sodium reacts with oxygen to produce sodium oxide.

Word equation: sodium + oxygen → sodium oxide

Formula equation: Na + O_2 → Na_2O

REMEMBER Note that copper and silver are non-diatomic elements so their symbols are written.

⊙ Example 2: Copper reacts with silver(I) nitrate to produce silver and copper(II) nitrate.

Word equation: copper + silver(I) oxide → silver + copper(II) nitrate

Formula equation: Cu + $AgNO_3$ → Ag + $Cu(NO_3)_2$

Balanced equations

REMEMBER The formulae in the balanced and unbalanced equations are the same. They should never be changed to balance an equation.

An equation is said to be balanced when the total number of atoms of each element on the left-hand side of the equation equals the total on the right.

Example 1:   When hydrogen reacts with chlorine, hydrogen chloride is formed.

Word equation: hydrogen + chlorine → hydrogen chloride

Unbalanced equation: H_2 + Cl_2 → HCl

There are more hydrogen and chlorine atoms on the left-hand side. Atoms can't just disappear. The 'missing' hydrogen and chlorine atoms must have formed another hydrogen chloride molecule, as this is the only product.

Balanced equation: $H_2 + Cl_2 \rightarrow 2HCl$

There are now the same number of atoms of each element on each side of the equation. In the balanced equation, numbers are put in front of the formula – never in between the symbols of the elements in a formula.

◎ *Try balancing the equations in examples 1 and 2 in the Chemical Equations section (page 88).*

State symbols

Symbols are sometimes shown in equations to indicate the states of the substances.

Example: $MgCO_3(s) + H_2SO_4(aq) \rightarrow MgSO_4(aq) + CO_2(g) + H_2O(l)$

(s) = solid, (l) = liquid, (g) = gas, (aq) = solution

☉ Spectator ions

Spectator ions are ions which don't take part in a chemical reaction but are present. This is best shown in an example.

Example: hydrochloric + sodium   →   water + sodium
                              acid   hydroxide   chloride

Formula equation: $HCl(aq)$ $+ NaOH(aq)$ $\rightarrow H_2O(l) + NaCl(aq)$

Ionic $H^+(aq) + Cl^-(aq) + Na^+(aq) + OH^-(aq) \rightarrow H_2O(l) + Na^+(aq) + Cl^-(aq)$
equation:

The only ions which have reacted to form a new substance are the $H^+(aq)$ and the $OH^-(aq)$. The other ions, $Na^+(aq)$ and $Cl^-(aq)$, are the spectator ions. The equation can be written leaving out spectator ions:

$H^+(aq) + OH^-(aq) \rightarrow H_2O(l)$

☉ Formula mass

The formula mass of a substance is obtained by adding together the relative atomic masses (RAM) of the atoms of the elements in the formula. (RAMs are found on page 4 of the SQA Data Booklet). The number of atoms of each element must be taken into account.

Example: magnesium chloride – $MgCl_2$

Element	RAM	No of Atoms	Mass
Mg	24.5	1	24.5
Cl	35.5	2	71.0
		Formula mass	95.5

The formula mass measured in grams (gram formula mass) is known as one mole. One mole of an element is simply its RAM measured in grams. Multiples and fractions of a mole can be calculated using:

BITESIZEchemistry

$$\text{moles} = \frac{\text{mass (g)}}{\text{mass of 1 mole}} \qquad \text{So mass} = \text{no. of moles} \times \text{mass of 1 mole}$$

The following table gives some examples.

How many moles?	What mass?
1. How many moles are in 77.5 g of sodium oxide (Na_2O)?	2. Calculate the mass of 0.5 moles of hydrogen sulphide (H_2S).
$$\text{moles} = \frac{\text{mass}}{\text{mass of 1 mole}}$$	$\text{mass} = \text{moles} \times \text{mass of 1 mole}$
$$= \frac{77.5}{62}$$	$= 0.25 \times 34$
$= 1.25 \text{ mol}$	$= 8.5 \text{ g}$

Calculations based on balanced equations

The numbers in front of formulae in balanced equations tell you the number of moles reacting and being produced. This allows masses of substance reacting and being produced to be calculated.

Example 1: $\qquad\qquad 2H_2 + O_2 \quad \rightarrow 2H_2O$

No of moles reacting: 2 mol 1 mol $\rightarrow$ 2 mol

Mass (g): $\qquad\qquad$ 4 g + 32 g $\rightarrow$ 36 g

Note: The total mass of reactants equals the total mass of products.

The balanced equation gives the proportions of reactants and products.

Example 2: $\quad$ Calculate the mass of hydrogen chloride produced when 10 g of hydrogen is completely reacted with chlorine.

Balanced equation:

	H_2	+	Cl_2	$\rightarrow$	2HCl
	1 mol		1 mol	$\rightarrow$	2 mol
	2 g	+	71 g	$\rightarrow$	73 g
so	10 g			$\rightarrow$	365 g

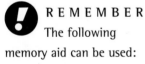
Concentration

The concentration of a solution depends on both the quantity of solute and the volume of solution. The unit of measurement is either grams per litre (g/l) or, more usually, moles per litre (mol/l).

The relationship between concentration (c), the number of moles of solute (m) and the volume of solution (v) is: $\qquad c = \dfrac{m}{v \text{ (in litres)}}$

90

Example 1: Calculate the concentration of a potassium hydroxide solution containing 0.25 mol of potassium hydroxide in 500 cm³ of solution.

m = 0.25 mol; c = 0.5 l

$$c = \frac{m}{v} = \frac{0.25}{0.5} = 0.5 \text{ mol/l}$$

Example 2: Calculate the concentration of a sodium nitrate solution containing 8.5 g of sodium nitrate in 250 cm³ of solution.

$$m = \text{mass / mass of 1 mole} = \frac{8.5}{85} = 0.1 \text{ mol}; \quad v = 0.25 \text{ l}$$

$$c = \frac{m}{v} = \frac{0.1}{0.25} = 0.04 \text{ mol/l}$$

Calculations from titrations

The volume of acid needed to neutralise an alkali (or vice versa) can be found by titration – you may have done this as a practical technique. The concentration of the alkali (or acid) can then be calculated as follows:

1. Use the volume of acid to calculate the number of moles of acid used.

2. Use the balanced equation for the reaction to calculate the number of moles of alkali reacted.

3. Use the number of moles of alkali to calculate the concentration of alkali.

Example 1: 30 cm³ of potassium hydroxide solution was neutralised by 20 cm³ of 0.1 mol/l sulphuric acid. Calculate the concentration of the potassium hydroxide solution.

1. $m_{acid} = c \times v = 0.1 \times 0.2 = 0.002 \text{ mol}$

2. $H_2SO_4 + 2KOH \rightarrow K_2SO_4 + 2H_2O$

   1 mol   2 mol

so, 0.002   0.004

3. $c_{alkali} = \frac{m_{alkali}}{v_{alkali}} = \frac{0.004}{0.03} = 0.13 \text{ mol/l}$

Alternatively, the concentration can be calculated using the relationship:

$$\frac{c_a \times v_a}{c_b \times v_b} = \frac{a}{b}$$

where: c_a = concentration of acid c_b = concentration of base (alkali)

     v_a = volume of acid v_b = volume of base (alkali)

     a = number of moles of acid, from balanced equation

     b = number of moles of base (alkali) from balanced equation

Example 1 can be calculated again using the method shown below, i.e. calculate the concentration of the potassium hydroxide solution (c_b).

Formulae, equations and calculations

Example 2:

$$\frac{c_a \times v_a}{c_b \times v_b} = \frac{a}{b}$$

so, $\quad c_b = \dfrac{c_a \times v_a \times b}{v_b \times a}$

$$= \frac{0.1 \times 20 \times 2}{30 \times 1}$$

$c_b = 0.133$ mol/l

● Percentage composition by mass

The percentage by mass of an element in a compound can be calculated as follows:

% by mass $= \dfrac{\text{mass of element in formula}}{\text{formula mass}} \times 100$

Example 1: Calculate the percentage, by mass, of nitrogen in ammonium nitrate (NH_4NO_3).

Mass of nitrogen in formula $\quad = (2 \times 14) = 28$

Formula Mass $= (28 + 4 + 48) = 80$

% nitrogen $\quad = \dfrac{28}{80} \times 100 = 35\%$

● Empirical formulae

The empirical formula gives the simplest ratio of atoms in a compound – ethane has the molecular formula C_2H_6 so its empirical formula is CH_3.

Empirical formulae can be calculated from either the mass or percentage of each element in a compound.

Setting out the calculation showing all the steps is important.

Example 1: An oxide of copper contains 3.175 g of copper and 0.4 g of oxygen. Work out the empirical formula of the oxide.

	Elements	Cu	O
divide by RAM	mass (g)	3.175	0.4
	number of moles	$\dfrac{3.175}{63.5} = 0.05$	$\dfrac{0.4}{16} = 0.025$
divide by smallest number	ratio	$\dfrac{0.05}{0.025} = 2$	$\dfrac{0.025}{0.025} = 1$

Empirical formula is Cu_2O

Example 2:   Work out the empirical formula of an oxide of iron, which was found to contain 70% iron and 30% oxygen.

	Elements	Fe	O
	%	70	30

% divided by RAM $\frac{70}{56} = 1.24$ $\frac{30}{16} = 1.88$

divide by smallest number

ratio $\frac{1.24}{1.24} = 1$ $\frac{1.88}{1.24} = 1.5$

multiply by 2 $1 \times 2 = 2$ $1.5 \times 2 = 3$

Empirical formula is Fe_2O_3

Practice questions

1   Work out the formulae of the following: a) magnesium b) hydrogen c) boron hydride d) carbon monoxide **G** e) copper(I) oxide f) cobalt(III) nitrate.

2   Write word and formulae equations for the following reactions:

   a) hydrogen reacting with bromine to form hydrogen bromide

   b) barium chloride and magnesium sulphate reacting to produce a precipitate of barium sulphate and magnesium chloride.

   G c) iron reacting with copper(II) nitrate solution to produce iron(II) nitrate and copper.

G 3   Balance the following equations:

   a) $Na + S \rightarrow Na_2S$

   b) $AgNO_3 + MgCl_2 \rightarrow AgCl + Mg(NO_3)_2$

G 4 a) Calculate the number of moles in 33 g of carbon dioxide (CO_2).

   b) Calculate the mass of 0.7 moles of magnesium nitrate, $Mg(NO_3)_2$

G 5   Calculate the mass of magneusium produced when 12.25 g of magnesium is completely burned in the air.
$2Mg + O_2 \rightarrow 2MgO$

G 6   Calculate the concentration of a calcium nitrate solution containing 0.1 mol of calcium nitrate in 200 cm³ of solution.

G 7   25 cm³ of sodium hydroxide solution was neutralised by 15 cm³ of 0.2 mol/l sulphuric acid. Calculate the concentration of the sodium hydroxide solution.
$2NaOH + H_2SO_4 \rightarrow Na_2SO_4 + 2H_2O$

G 8   Calculate the percentage, by mass, of nitrogen in ammonium sulphate, $(NH_4)_2 SO_4$

G 9   Work out the empirical formula for the compound which was found to contain 0.2 g of hydrogen and 3.2 g of oxygen.

G 10   A hydrocarbon contains 80% carbon. Work out the emperical formula of the hydrocarbon.

Answers

Page 11

1. a) Rise in temperature.
   b) Use of a thermometer.

2. C and D

3. Salt is the solute, water is the solvent.

4. a) Colour changes.
   b) Magnesium copper(II) sulphate $\rightarrow$ magnesium sulphate + copper
   c) Magnesium and copper

page 13

1. a) The flask gets very hot.
   b) Hydrogen gas is given off.

2. a) Speeds up the reactions
   b) 1 g (no change)
   c) $35^\circ C$, $100\ cm^3$ of $0.1mol/l$ hydrogen peroxide, 1 g manganese dioxide

page 15

1. a) Any group 1 metal
   b) platinum or rhodium
   c) fluorine
   d) argon
   e) beryllium
   f) magnesium
   g) mercury

2. a) A and F
   b) B
   c) A and C

3. a) B
   b) F

page 19

1. a) There are equal numbers of positive and negative charges which balance each other.
   b) There are more neutrons than protons.

2. a)

Type atom	Number of protons	Number of neutrons
$^{28}_{14}Si$	14	14
$^{29}_{14}Si$	14	15
$^{30}_{14}Si$	14	16

   b) Isotopes

page 23

   a) C
   b) F
   c) E

page 25

1. A and D

2. a) conductor
   b) zinc
   c) Ions are not free to move (powder = solid).

3. a) B and F
   b) D
   c) E

page 27

1. a) Electrolysis
   b) Chlorine
   c) Cu^{2+} ions attracted and gain two electrons to form Cu atoms (metal).
   d) At the positive eletrode $2Cl^-(aq) \rightarrow Cl_2(g) + 2e^-$; at the negative electrode $Cu^{2+}(l) + 2e^- \rightarrow 2Cu$

2. a) The breaking up of a compound by passing electricity through it.
   b) Ions are not free to move.
   c) An electrode may be touching the metal crucible.
   d) At the positive electrode: $2Cl^-(l) \rightarrow Cl_2(g) + 2e^-$; at the negative electrode: $Li(l) + e^- \rightarrow Li(l)$

page 31

1. a) Formed over millions of years from dead plants which were subjected to heat, pressure and bacterialogical action.
   b) Oil or natural gas.

2. a) Exothermic
   b) Carbon dioxide + water

3. a) Oxygen
   b) Carbon dioxide + water
   c) To condense the water formed.

page 33

1. a) A and B
   b) B

2. a) Gasoline
   b) Road building

   c) Gasoline
   d) Bigger molecules
   e) To produce smaller, more useful fractions.

page 35

1. a) Carbon monoxide
   b) Not enough oxygen for complete combustion.

3. a) Nitrogen and carbon dioxide
   b) Platinum or rhodium
   c) Bigger surface for chemicals to react on.
   d) Increase the air/fuel ratio.

page 41

   a) Lemon juice and vinegar
   b) Milk of magnesia and ammonia solution

page 45

1. a) B and E
   b) B and E

2. a) Neutralisation
   b) No more gas given off.
   c) To remove unreacted carbonate.
   d) Copper does not react with acid.

3. a) B and F
   b) C

page 51

1. a) (i) Ions can't move in a dry powder (solid).
       (ii) Chemicals are used up.
   b) Rechargeable.

2. a) C
   b) A and F

page 55

1. a) An alloy
   b) Heat conduction

2. E

3.

Metal added to iron	Use of steel
titanium	rockets
cobalt	magnets
manganese	railway lines
tungsten	drill bits

BITESIZEchemistry

4 a) Lightweight
  b) Strength

page 57

  a) E
  b) F
  c) C

page 59

1 a) D
  b) C
  c) D

2 Sodium and potassium are so reactive they react quickly with oxygen and water in the air.

3 a) R, Q, S, P
  b) R: any metals above Mg in the reactivity series
  Q: Mg
  S: iron, tin or lead
  P: any metal below lead in the reactivity series.

page 61

1 a) Stops the oxygen and water attacking the metal.
  b) Sacrificial protection or connect to the negative terminal of a power supply.
  c) Sea water contains ions which increases speed of corrosion.

2 a) (i) No oxygen present
  (ii) Sea water contains ions.

3 a) Fe^{3+} (aq)
  b) Ferroxyl indicator
  c) Green to blue.

page 63

1 a) Water and oxygen
  b) Mg is higher than Fe in the electrochemical series so electrons flow from Mg to Fe. Cu is lower than Fe so electrons flow from Fe to Cu.

2 C and D

page 69

1 a) Monomers with C=C double bonds add together, breaking the double bond.

b)

```
    H   CH3   H   CH3   H   CH3
    |    |    |    |    |    |
 -  C  - C  - C  - C  - C  - C  -
    |    |    |    |    |    |
    H    H    H    H    H    H
```

  c) Cracking oil.

2 a) polymethyl methacrylate

b)

```
   H    CH3    H    CH3    H    CH3
   |     |     |     |     |     |
 - C  -  C  -  C  -  C  -  C  -  C -
   |     |     |     |     |     |
   H   COOCH3  H   COOCH3 H   COOCH3
```

page 71

1 a) C and D
  b) Heat and electrical insulator

2 a) (i) polyacrylonitrile
  (ii) Thermoplastic
  b) They are toxic (poisonous).
  (c) Hydrogen cyanide (and carbon monoxide)

page 75

1 a) Nitrogen, phosphorous and potassium.
  b) Root nodules contain bacteria which converts nitrogen from the air.
  c) From soluble nitrogen compounds in the soil.
  d) The world's population is growing quickly.

2 a) Soluble and contains a high percentage of nitrogen.
  b) They can be washed into our drinking water. They can pollute rivers and lochs causing fish to die.

page 81

1 a) carbon dioxide in; oxygen out
  b) carbohydrates
  c) traps light

2 Plants use up carbon dioxide produced by animals. Plants produce oxygen used up by animals.

3 a) E
  b) B

page 83

1 a) Carbohydrates
  b) Enzymes
  c) Add Benedict's or Fehling's solution and warm the mixture. If the Benedict's colour changes from blue to orange/brown, then glucose is present.

2 a) D and E
  b) A and B
  c) D

page 93

1 a) Mg
  b) H_2
  c) BH_3
  d) CO
  e) Cu_2O
  f) $Co(NO_3)_3$

2 a) hydrogen + bromine $\rightarrow$ hydrogen bromide
  $H_2 + Br_2 \rightarrow HBr$
  b) barium chloride + magnesium sulphate $\rightarrow$ barium sulphate + magnesium chloride
  $BaCl_2 + MgSO_4 \rightarrow BaSO_4 + MgCl_2$
  c) Iron + copper(II) nitrate $\rightarrow$ iron(II) nitrate + copper
  $Fe + Cu(NO_3)_2 \rightarrow Fe(NO_3)_2 + Cu$

3 a) $2Na + S \rightarrow Na_2S$
  b) $2AgNO_3 + MgCl_2 \rightarrow 2AgCl + Mg(NO_3)_2$

4 a) 0.75 mol
  b) 104 g

5 20.25 g

6 0.5 mol/l

7 0.24 mol/l

8 21.2%

9 HO

10 CH_3

Index

acid   38, 40, 42, 43
acid rain   38, 44, 72
addition reaction   29, 36
alcohol (drinks)   78, 85
alkali   38, 40, 42
alkali metals (group 1)   14
alkanes   28, 36, 64
alkanol   78
alkenes   28, 36
alloy   53, 54
ammonia   72, 76
ammonium salts   75
aqueous (aq)   14
atom   16
atomic number   16
battery   46, 62
Benedict's test   78, 83
biodegradable   64, 70
blast furnace   52, 56
bond
    covalent   20, 22
    ionic   20, 22
brass   53, 55
burning   29, 34, 44
calculations
    from equations   90
    from volumetric titrations   91
carbohydrate   78, 80, 82
carbon   52
    dioxide   28, 35, 52, 56
    monoxide   34, 35, 52, 56
catalyst   8, 13, 37, 72, 77, 79, 85
catalytic converter   29, 35
cell   48, 49
chemical reaction   8, 10
coal   30, 56
combustion   29
compost   72
compound   9, 10
concentration   90
conductivity   20, 24, 52, 54
conductor   24
corrosion   52, 53, 60
covalent network   20, 24
cracking   28, 36, 66
current   21
cycloalkanes   28, 36, 37
decomposition   52
density   55
diatomic molecule   20, 24
digestion   79, 84
dilution   41
direct current (d.c.)   20, 21, 26
disaccharide   78, 82
displacement   46, 49
electrochemical series   46, 49
electrolysis   20, 26, 52
electrolyte   46, 48, 61

electron   16, 46, 56, 67
electroplating   53
element   8, 14
empirical formula   92
enzyme   13, 78, 84, 85
equation   88
    balancing   88
    calculations from   90
    chemical   88
    word   10
ethanol   78, 85
ethane   92
exothermic   73, 79
fermentation   78, 85
ferroxyl indicator   52, 53, 60
fertiliser   72, 74, 75
finite resources   30, 57
flammable   32
formula   86
    elements   87
    mass   89
fossil fuel   28, 30, 78
fraction   28, 32
fractional distillation   28, 32
fuel   28, 30, 34
glucose   78
group   14
Haber Process   72, 76
homologous series   28, 37
hydrocarbon   28, 30, 34, 36
    saturated   36
    unsaturated   36
hydrogen   29, 30
hydrolysis   78, 79, 85
hydroxide ion   53, 62
indigestion   38, 40, 42
insulator   70
ion   20, 24, 41
ion (salt) bridge   46, 49
ion–electron equation   26, 50
ionic lattice   20, 22
iron   52, 56, 62
isomer   29, 37
isotope   9, 18
leguminous   73, 74
manure   72, 73
mass number   17
metal
    extraction   52, 53, 56
    physical properties   52, 54
    reaction with acids   58
    reaction with oxygen   58
    reaction with water   58
methane   30
mole   90, 91
molecule   20
monomer   66, 67, 84
monosaccharide   78, 82

natural gas   31
neutralisation   38
neutron   9, 16
nitrate   75
nitric acid   72, 75, 77
nitrogen cycle   74
nitrogen-fixing bacteria   74
nitrogen oxide   77
noble gases   14
non-metal oxides   38, 40
nucleus   9
nutrients   72
oil (crude)   28, 31
ore   52, 56, 58
Ostwald Process   72, 77
oxidation   46, 50, 60
percentage composition   92
periodic table   8, 14, 15
petrol   29, 32, 34, 44, 72
pH   38, 40
photosynthesis   78, 79, 80, 81
pollution   28, 34, 64, 72, 75
polymer   64, 66, 67, 84
    addition   64, 69
    condensation   78, 84
polymerisation   64, 69
polysaccharide   78, 82
population   72, 74
precipitate   44
proton   9, 16
rainforest   81
reactivity series   58
rechargeable cells   46
recycle   70, 74
redox   46, 50
reduction   46, 50, 57, 60
relative atomic mass   19
respiration   80, 81
rusting   52, 53
sacrificial protection   62
salts   42
    soluble   38, 44
    insoluble   38, 44
solubility   25
solution   11, 41
solvent   11
speed (rate) of reaction   12
starch   78, 84
state   8, 89
sucrose   82
sulphur dioxide   28, 34
synthetic   64, 71
toxic   70
transition metals   14
universal indicator   40
valency   86